Monographs of the Palaeontographical Society

The Palaeontographical Society was established in 1847, and is the oldest Society devoted to study of palaeontology worldwide. Its primary role is to promote the description and illustration of the British fossil flora and fauna, via publication of an authoritative monograph series. These monographs cover a wide range of taxonomic groups, from microfossils, trilobites and ammonites through to Coal Measure plants, mammals and reptiles, and from all ages from Cambrian to Pleistocene. They form a benchmark for understanding the past life of the British Isles and many include the original descriptions of numerous key species. The first monograph (on the Crag Mollusca) was published in March 1848 and the Society still continues this work today. Notable authors in the series include Charles Darwin (fossil barnacles) and Richard Owen (dinosaurs and other extinct reptiles). Beginning in 2014, the Cambridge Library Collection and the Society are collaborating to reissue the earlier publications, focusing on monographs completed between 1848 and 1918.

A Monograph of the Crag Mollusca

The Pliocene–Pleistocene Crags of East Anglia are an incredibly rich source of fossil shells, many belonging to extant Boreal and Mediterranean genera. Dominated by marine gastropods and bivalves, the deposits also contain evidence of terrestrial and non-marine gastropods and bivalves, brachiopods, and extensive epifauna including bryozoans. Published between 1848 and 1879 in four volumes, the latter two being supplements with further descriptions and geological notes, this monograph by Searles Valentine Wood (1798–1880) covers more than 650 species and varieties of fossil mollusc. For each species Wood gives a synonymy, diagnosis (in Latin), full description, dimensions, occurrence and remarks. The supplements also provide a breakdown of the species and their current distribution. The detailed plates were prepared by the conchologist George Brettingham Sowerby and his namesake son. Volume 1 (1848) covers gastropods and scaphopods, illustrated in 21 plates.

A Monograph of the Crag Mollusca

Or, Descriptions of Shells from the Middle and Upper Tertiaries of the East of England

VOLUME 1: UNIVALVES

SEARLES V. WOOD

CAMBRIDGE
UNIVERSITY PRESS

CAMBRIDGE
UNIVERSITY PRESS

University Printing House, Cambridge, CB2 8BS, United Kingdom

Cambridge University Press is part of the University of Cambridge.
It furthers the University's mission by disseminating knowledge in the pursuit of
education, learning and research at the highest international levels of excellence.

www.cambridge.org
Information on this title: www.cambridge.org/9781108076883

© in this compilation Cambridge University Press 2014

This edition first published 1848
This digitally printed version 2014

ISBN 978-1-108-07688-3 Paperback

THE

PALÆONTOGRAPHICAL SOCIETY.

INSTITUTED MDCCCXLVII.

LONDON:

MDCCCXLVIII.

A MONOGRAPH

OF

THE CRAG MOLLUSCA,

OR,

DESCRIPTIONS OF SHELLS

FROM THE

MIDDLE AND UPPER TERTIARIES OF THE EAST OF ENGLAND.

BY

SEARLES V. WOOD, F.G.S.

PART I.

UNIVALVES.

LONDON:
PRINTED FOR THE PALÆONTOGRAPHICAL SOCIETY.
1848.

C. AND J. ADLARD, PRINTERS, BARTHOLOMEW CLOSE.

INTRODUCTION.

A GEOLOGICAL HISTORY or details of the different Formations, not coming within the object of the Palæontographical Society, it is only necessary here to state that this work may be considered as an illustrated description of my Catalogue of Crag Shells, published in the 'Annals and Magazine of Natural History,' 1840-42, including such Species as have since been discovered, in addition to those already enumerated.

It is intended to include as *Crag Shells* all those species, hitherto considered as coming under this general and well-understood denomination, from the *three* different Periods, into which that Formation has been divided by geologists, viz., the Coralline Crag, as representing in this country the remains of the Miocene Period; the Red Crag, those of the Pliocene Period; and the Mammaliferous Crag, those of the Pleistocene Period. In the latter are included the several species that have been obtained at Bridlington, as that bed is now considered to be a Marine Formation of the Pleistocene Period, and consequently synchronous with the Estuary deposits of Bramerton and Thorpe. The northern beds of the Clyde, &c., may belong to the same Period; but I am not sufficiently acquainted with those deposits to express an opinion on that point, as I have seen only such shells from them as may be considered identical with existing species. A Catalogue of the Mollusca from the Lacustrine or Fluviatile deposits of Grays,* Clacton,† Stutton,‡ and Copford,§ will be given as an Appendix to the Second Part of this work, as those beds are

---

* Grays, on the river Thames, twenty-one miles from London, in the county of Essex.
† Clacton, in Essex, on the coast, seven miles south-west of Walton Naze.
‡ Stutton, in Suffolk, on the banks of the river Stour, six miles south of Ipswich.
§ Copford, in Essex, about four miles south-west of Colchester.

probably the Freshwater equivalents to the Crag Periods; the first three mentioned localities belonging, perhaps, to the Red Crag Period, while the Copford deposit may be of a more modern date. As these localities contain Mollusca, of which all the species are identical with existing forms, and which have been already figured and described, it is presumed that a name alone, with reference to the work in which each species is given, will be sufficient for geological purposes. The land and Freshwater shells delineated here, having been found in the Red and Mammaliferous deposits, intermixed with Marine species, may be considered as forming an integral portion of these beds, as *their* geological age cannot be doubted: they therefore cannot well be omitted, more especially as they exhibit, in most instances, a variation in form, more or less considerable, from their typical characters, thus exciting our interest regarding the climatal and other conditions under which they existed, and which appear to have exerted no little influence upon most of the animals of that Period.

In justice to my predecessors in this field of geological and palæontological research, a small space may be allotted for a brief summary of their labours.

'An Attempt towards a Natural History of Fossils,' by John Woodward, in 1729, contains the first notice of any organic remains from the Crag Formation.

Robert Dale, in his 'History and Antiquities of Harwich,' published in 1730, gives copious descriptions, as well as representations, of several species from the Crag, that were found at that time in the Cliff near Harwich; the execution of some of these figures is by no means to be despised, and the shells may be readily recognised. This author states that during the time he had observed the Cliff, the wearing away of its materials was so rapid, that after a period of forty years but little of the stratum remained from which he had, in the early part of his life, collected his fossils; and although portions of thc Crag were remaining during the time of his publication, all vestiges of it have long disappeared from that locality.

Parkinson, in his 'Organic Remains of a Former World,' 1811, gives a few good figures of Crag shells.

In 1816, William Smith, "the Father of Geological Science in England," published a plate of good figures of Crag fossils in his 'Strata Identified by Organized Fossils.'

Mr. Richard Cowling Taylor published, in the 'Philosophical Magazine,' 1822-24, some observations on the Norfolk Cliffs, as well as a Paper, in the 'Geological Transactions,' upon the Tertiary deposits of Norfolk and Suffolk; some good figures of Crag Corals were also given by him in the 'Magazine of Natural History,' 1830.

In 1826, Mr. Robberds published some 'Observations on the Eastern Valleys of Norfolk.'

In 1833, 'An Outline of the Geology of Norfolk' was published by Samuel Woodward the elder, in which there are some good figures of Crag fossils, from the beds in the immediate neighbourhood of Norwich, relating only to the Mammaliferous Crag.

The geological features of the upper Tertiaries of this country, more especially those portions from which the shells here illustrated have been extracted, have been ably delineated by Mr. Lyell, in his 'Principles' as well as in his 'Elements of Geology;' and also in a special Memoir, published in the 'Magazine of Natural History' for 1839.

Mr. Charlesworth also published a series of Papers in the 'Philosophical Magazine' for 1835, in which the Crag Formation was separated into *three* different Periods, the oldest of which he designated the "Coralline Crag," as expressive of its contents, that Formation being in some places composed entirely of Zoophytic remains. The succeeding Period, or "Red Crag," was so denominated from its peculiar ochreous colour, the beds being strongly stained by large quantities of the hydrous oxide of iron; but it was considered as geologically distinct, on account of the absence of the great mass of Zoophytes peculiar to the lower beds, as well as of the introduction of a new group of Testacea. A further examination of what had previously been considered as an extension of the Red Crag Formation into Norfolk, induced that author to suspect the Crag of Norfolk to be of posterior age, and, as position in this case was no guide to such determination, he appealed to the certain evidence of zoological distinction, which presented a more recent aspect than either of the preceding; and as amongst its imbedded fossils are the remains of numerous Mammalia, he gave to this Formation the title of "Mammaliferous Crag." These terms are employed in the present work.

In 1836, Mr. John Morris published some observations, in the 'Magazine of Natural

History,' upon the Freshwater Formation of the upper Tertiaries at Maidstone, in Kent. The Period to which this bed belongs has not yet been satisfactorily determined.

In the eighth volume of the 'Magazine of Natural History,' 1839, Mr. W. Bean gave an account of a deposit at Bridlington Quay, whence some of the shells I have figured were obtained.

In 1840, Mr. John Brown, of Stanway, communicated a Paper to the 'Magazine of Natural History,' describing a Lacustrine deposit at Clacton, on the coast of Essex, which is probably the Freshwater equivalent to the Red Crag Formation. This gentleman also printed and privately distributed a list of Crag Shells, obtained by him from a place called Beaumont, in Essex. In this list, however, there are not the names of any species but such as have been obtained at Walton-on-the-Naze by myself; and as I have not been able personally to inspect the locality referred to by Mr. Brown, no special reference has been made to it.

In December, 1843, Professor Henslow read a Paper before the Geological Society, describing fossils from the Crag at Felixstow, which he considered to be the Croprolitic remains of whales; and introduced a notice respecting some tympanic bones, which were referred to the genus Balænodon by Professor Owen.

As early as the year 1812, Mr. James Sowerby, senior, commenced the 'Mineral Conchology,' which for a long series of years has been almost the only publication by which British fossil shells of all Periods have been made known to the public in this country; and this work has been ably continued by his son, Mr. James De Carle Sowerby, to the present time.

The first portion of the present work contains descriptions of univalve shells, or the calcareous remains of Gasteropodous Molluscs. The great variety of forms presented by the testaceous coverings of this class of animals, depends upon the height or length of the cone. This ranges through every degree of angularity, from the nearly discoidal form exhibited by some of the Patelliform species, in which the cone is so depressed as to form an angle, from the vertex to the margin, of 170 degrees, to that of an elongated tube, extended in some cases so much as to become nearly cylindrical, and twisted into almost every conceivable form of spiral, for the convenience, or rather according to the necessities of the animal. Among the spiral shells these varying forms are the result of an obliquity in the mode of growth, caused by a greater development or increase of one side of the opening, whereby an heliciform direction is given to the shell in the process of enlargement. It is, perhaps, needless to say, that the mode of increase or growth of the shell is by the successive deposition of calcareous matter secreted by the mantle, and deposited on the margin of the aperture, and moulded as it were upon the soft body of the animal, by which means the impress of every peculiarity is formed, and permanently fixed in the calcareous covering. It may, however, be observed, that some of the exterior ornaments of these shells are formed by an extension or protrusion of the mantle, whereby a fimbriated, and sometimes a spinose, varix is formed upon the margin of the outer lip, producing many varied and beautiful ornaments upon the spiral shell, which, however, have no corresponding characters upon the soft body of the animal. It is from these appearances alone, that presumed specific distinctions can be determined; and the characters drawn from them, although they may be considered as somewhat extrinsic to the animal, are the only means afforded to the palæontologist for the determination of species. From the recent researches of Mr. Bowerbank and Dr. Carpenter, in their Microscopic Examinations into the Structure of Shells, it would appear that there is a more intimate connexion between the animal and its envelope during its lifetime than could be imagined by the supposition of a simple addition of earthy matter deposited upon its cuticle; and that this strong covering is an organized body, and part and parcel of the animal itself. Important as are those observations and suggestions, which bear considerably on generic groups, and by which, perhaps, even minor distinctions may hereafter be pointed out, the specific determinations referred to in the present monograph have not been submitted to such tests, but are merely drawn from the various external differences existing in the shells, and which variations in structure are generally considered as resulting from some internal economy of the animal.

This mode of determination may, perhaps, be correctly applicable only to shells in perfect condition; whereas, some of those figured in this work have undergone considerable alteration, either by the loss of colour and animal matter, or by decomposition of the outer coating, while many of them have sustained mutilation by the mechanical action of the waves. These accumulated difficulties are serious impediments to accurate comparisons; and the identifications in many cases must be looked upon only as approximations: a correct Marine Fauna of this period will, it is to be feared, be for many years to come a desideratum.

In regard to the admeasurement of shells in the present work, minute fractional descriptions have not been very particularly attended to, but a general approximation only has been given, from a belief that a comparison of the dimensions of the aperture and the axis of a shell is a test of fallacious dependence. In many instances, specimens of the same species have a much greater range in variation than, as a general character, would be exhibited between two proximate species; the aperture in some individuals being equal to two thirds of the length of axis, while in others of the same species it does not exceed one third. In an elevated or cylindrically formed shell, with numerous volutions, the aperture of a young individual bears a greater proportion to the axis of the shell than it does when the animal is fully grown; and it is not always possible to ascertain when an individual has reached a state of maturity. The form of the aperture is nearly the same at all ages or states of the individual, but the lengthening or shortening of the convoluted cone, which may be the result of external causes, will materially affect any mathematical proportion that may be assumed as a standard for specific determination; and the spiral angle, or angle of volution,* upon which some Continental conchologists have placed so great reliance, is a test, in my opinion, equally illusory and deceptive. In *Littorina littorea*, for example, the variations extend, in what I firmly believe to be individuals of the same species, from a subulate or tapering form of volution, denoting an angle of scarcely twenty degrees, to another form which, by the application of the same rule, would show an angle of ninety degrees; this may be seen in Plate X, where the figures are by no means exaggerated, but rather fall short of the two extremes. A like want of permanent regularity in proportionate dimensions is equally exhibited in the Acephala or bivalve molluscs; and minute attention to accurate detail or mathematical proportions is rarely to be depended upon, even in that class. In some species, of which the general form may be considered as transverse, the character may be traced through trifling and almost imperceptible gradations until it is decidedly elongate, so that mathematical proportions in such cases are of no determinate value. This is more particularly evident in some of the species in the Crag Formations, where, perhaps, a rather more than ordinary degree of variation may be seen; and such are the extraordinary varieties in many of the species, that no reliance can be placed, for specific determination, upon minute proportionate dimensions, although, as an auxiliary, they may in some cases be usefully referred to.

No alteration has been thought necessary in the generally accepted terms of

---

* The angle formed by a line drawn from the vertex on each side of the shell, which varies, of course, upon the greater or less depression of the volution, the sutural angle, also, depending upon the same cause.

longitudinal and *transverse,* applying them to what is so in appearance only; and the term *labrum* is applied to that portion of the peritreme which corresponds with the right hand of the investigator; while *labium* is given to its opposite or columella lip.

For the dates of publication of the different genera, I am much indebted to the work of M. A. N. Hermansen, entitled 'Indicis Genera Malacozoorum Primordia,' now in course of publication; and also to a very valuable Paper on a similar subject, by J. E. Gray, Esq., published in the 'Proceedings of the Zoological Society,' November, 1847.

Considerable difficulty has been experienced in selecting an Order of Arrangement for the shells described. An immense mass of information regarding the soft parts of Mollusca has been obtained within the last few years; and the animal inhabitants of no less than five thousand species are now known, and have been examined;* yet, notwithstanding this additional information, we are still without a Natural Classification, to which anything like a general concurrence appears to be given; and it is only necessary to compare such as have been most recently published on the Continent with those of our own naturalists in England, to observe the great dissimilarity between them. It has therefore been thought most advisable to employ the old artificial arrangement, with some slight modifications, uniting in different sections those shells in which the aperture is emarginate, or furnished with a canal for the siphon of the animal, and those in which the mouth is circular, or with a continuous peritreme, and merely placing in advance the Pulmonata, or air-breathing animals, after the manner of Cuvier, conceiving them to be more highly organized than those which breathe simply by means of gills.

If at first sight it should be considered improbable that some of the extreme varieties in form belonged to one species, it must be stated that these results have been arrived at, not only by the possession, but by the careful examination of a large series of individuals.

Some information has been introduced that may be thought unnecessary; but I have considered that the newly-established Palæontographical Society will, in all probability, contain among its members students just entering the field of Palæonto-

---

* Gray, Zoological Proceedings, p. 132, November, 1847.

logical research, to whom every species of information, although familiar to those long engaged in the study, will be acceptable.

It has been, of course, my endeavour to make the work as complete as possible, though it merely professes to be a plain and faithful delineation and history of the " Remains of Mollusca from the Crag." I do not pretend to put it forth as free from errors; on the contrary, it would be surprising if they were not *numerous*, more especially when it is recollected that in many instances, in consequence of the insufficient materials, and imperfect and altered condition of the shells, a more than ordinary degree of labour is imposed upon the Palæontologist.

The figures, in every instance, have been engraved from specimens in my own cabinet, and obtained by myself, excepting where otherwise particularly expressed ; and I beg to return my sincere thanks to my friends for their liberal assistance during the progress of the work ; particularly to Miss Alexander, Mr. Lyell, Capt. Alexander, Mr. Gibson, Mr. Charlesworth, Mr. Daniel, Mr. Perry, Mr. Robert Fitch, and Mr. Wigham, who have allowed me the use of their best specimens for the purpose of comparison and for figuring ; and to Mr. Bean and Mr. Leckenby, for the loan of their specimens from the Bridlington beds. I beg also to express my obligations to several friends who have assisted me with their opinions upon conchological subjects, especially to Professor E. Forbes, Mr. Alder, Mr. Hanley, and Mr. W. Thompson.

SEARLES V. WOOD.

January, 1848.

A MONOGRAPH

OF THE

MOLLUSCA FROM THE CRAG.

CLASS—CEPHALOPODA.

No remains of any animal belonging to this Class have, that I am aware of, been detected in any of the three Formations into which the Crag is divided. My cabinet contains a few specimens of Belemnites belonging to two different species, found by myself among the gravelly portion of the Red Crag. These, however, as they have been introduced from the older rocks, and have no connexion with the Tertiary Periods, do not require to be here particularised.

CLASS—GASTEROPODA.

ORD. *PULMONATA*, Cuv.

TERRICOLA, FLEM.

HELIX,* *Linn.* 1758.

Generic Character. Shell turbinated, orbicular, subglobose or depressed, light and thin : spire slightly elevated, with several more or less convex volutions, mostly smooth ; last whorl often large and ventricose, with an oblique aperture of a lunate or semi-ovate form, generally broader than long, impressed by the prominent part of the body whorl : peristome confluent with the columella, mostly strengthened internally by a thickened rib, and often having a reflected edge, especially on the left side covering the umbilicus, without an operculum.

Animals belonging to this order (Pulmonata), differ from the generality of the Gasteropoda in being capable of respiration direct from the atmosphere, and are fur-

---

* Etym. Ἕλιξ, *quod contortum est.*

1

nished with a simple pulmonary cavity, situated beneath the edge of the mantle, on the right side of the animal ; most of them are covered and protected by a calcareous shell, more or less convolute. Those which are constant inhabitants of the water, come occasionally to the surface for a supply of air, a necessary and vital element, which they cannot sufficiently obtain from the water alone.

As the species of this genus are entirely terrestrial, the presence of specimens found associated with marine shells in the Mammaliferous Crag may be due to the transporting power of rains and floods, conveying them into the ancient estuary, and to their buoyancy (more especially if dead shells) permitting them to be carried to greater or less distances, before they become entombed with the marine exuviæ.

Upwards of a hundred synonyms are given by Hermansen to this genus as originally established, many authors conceiving a very slight alteration of character sufficient to constitute generic distinction.

A genus like the present, which, even when shorn of all those marine, fluviatile, and terrestrial forms that were included in it by Linnæus and other ancient authors, contains in a living state, nearly one thousand species, might, however, be considered as deserving of subdivision, were it only to assist the student in the difficult task of investigation.

1. HELIX HISPIDA.   *Linn.* Tab. I, fig. 3, *a—c.*

HELIX HISPIDA.   *Lin.* Syst. Nat. p. 1244, 1766.

—   *Müller.* Verm. 11, p. 73, No. 268, 1774.

—   *Mont.* Test. Brit. p. 423, t. 23, fig. 3, 1803.

—   *Rössmasler.* Iconog. t. 31, fig. 426-427, 1835.

—   *Gray.* 2d edit. of Turt. Man. p. 154, t. 4, fig. 51, 1844.

—   *S. Wood.* Catalogue 1842.

—   var. concinna.   *Jeffreys,* Synopsis.

H. Testá orbiculato-convexá; spirá depressá, anfractibus 4—5 *vix carinatis; aperturá semilunari, labro tenui, sub-reflexa ; umbilico profundo.*

Shell orbiculato-convex, with a somewhat depressed spire ; whorls four or five, slightly carinated ; aperture semilunate, with a moderately sized, and deep umbilicus.

Diameter, $\frac{1}{4}$ of an inch ; *elevation,* $\frac{1}{7}$ ditto.

Locality. Mam. Crag, Bulcham.   Recent, Britain.

Several specimens have been kindly lent to me for comparison, from these, with a few of my own, I have no hesitation in assigning this to *H. hispida.* Two varieties are among those belonging to Mr. Lyell, one of which has the name of *H. plebeium* upon the tablet, which I think, however, is only a variety of *H. hispida* with a spire rather more depressed, and an umbilicus a little larger.   I have the same varieties from Stutton and Clacton.   This shell appears to present a slightly carinated form in the young state, though not quite so much so in the middle of the volution as in *H. rufescens ;* this character is less distinct in the older specimens.

2. HELIX PULCHELLA. *Müll.* Tab. I, fig. 4, *a—c.*

HELIX PULCHELLA. *Müll.* Verm. p. 30, No. 232, 1773.
— PALUDOSA. *Mont.* Test. Brit. p. 440, 1803.
ZURAMA PULCHELLA. *Leach.* Moll. p. 108, 1819.
LUCENA PULCHELLA. *Hartman.* t. 1, fig. 6, 1821.
HELIX COSTATA. *Flem.* Brit. An. p. 263, 1828.
AMPLEXIS PALUDOSIS. *Brown.* Con. Illust. pl. 41, f. 76, 77, 1827.
— CRENELLUS. - - - pl. 41, f. 78, 79.
HELIX MINUTUS. *Say.* Journ. Acad. Nat. Sc. I. 123, 1830.
— PULCHELLA. *Gray.* 2d edit. of Turt. Man. p. 141, pl. 5, fig. 49, 1844.
— — *S. Wood.* Catalogue 1842.

H. Testá minutá, orbiculato-depressá, umbilicatá; anfractibus 4, convexis, suturis depressis; labri margine crasso, albo, reflexo; umbilico magno.

Shell small, orbiculato-depressed, with about four volutions, finely marked with lines of growth; suture deep; aperture subcircular; peritreme thickened and reflected; a large and open umbilicus, showing the volutions within.

Diameter, $\frac{1}{10}$ of an inch; *elevation,* half the diameter.

Locality. Red Crag, Bawdsey.   Recent, Britain.

The only specimen of this species in my cabinet was found in loose and disturbed Crag; it is, however, deeply tinged with the hydrous oxide of iron, having all the appearance of a genuine Crag shell; and as it is very abundant in the fluviatile deposits of Stutton and Clacton, it was in all probability a living species during the Red Crag period.

3. HELIX ARBUSTORUM (?) *Linn.* Tab. I, fig. 2.

HELIX ARBUSTORUM (?) *Linn.* Syst. Nat. p. 1245.
— *Draparnaud.* Moll. p. 38, t. 5, fig. 18, 1805.
— Turt. Man. pl. 3, fig. 25, 1819.
— *Rössmasler.* Iconog. p. 56, pl. 1, fig. 4, 1835.

H. Testá subglobosá, perforatá, lævigata (?); spirá obtusá; anfractibus quinque, subcarinatis; aperturá oblique semilunari, labri margine reflexo; umbilico obtecto.

Shell subglobose and perforated; smooth (?), with a short conoidal spire; volutions five, slightly convex above, and rather compressed beneath, very obtusely carinate; margin reflected, covering the umbilicus.

Diameter, $\frac{3}{4}$ of an inch; *elevation,* $\frac{1}{2}$ ditto.

Locality. Mam. Crag, Southwold.   Recent, Britain.

The specimen figured is from the cabinet of Mr. Lyell, and is the only one I have seen. I have assigned it to the well-known recent species, although it is considerably altered in appearance, and differs from the general form of that shell in being more depressed, and in the obtusely or roundedly carinated form of the volutions. I have, however, some recent specimens as much depressed, and some which, in the young state, show a carinated form of volution. It somewhat resembles *H. Turonensis,* but that species is generally much larger, and the peristome of our shell is never so much reflected or thickened, more especially at the lower and inner part of the mouth.   The perfectly smooth appearance of the specimen is in all probability the result of attrition.

A recent specimen from Ayrshire, much depressed in form, has been obligingly sent me by Wm. Thompson, Esq., of Belfast, which very strongly resembles our Crag shell.

4. HELIX RYSA.   *S. Wood.* 1847. Tab. I, fig. 1.

H. Testá orbiculato-depressá, perforatá, supra convexá, radiatá, rugosá, seu corrugatá; subtus rotundatá,' profundè umbilicatá et subtilissimè striatá; anfractibus septem obtuse carinatis, carinis prope suturam distinctis: aperturá subdepressá, latè lunari; peritremate acuto, reflexo.

Shell orbiculato-depressed, perforated, convex, rugosely striated, or rather corru gated, on the upper half of the volutions, rounded, and faintly striated on the under surface, with a large and deep umbilicus; volutions seven, obtusely but distinctly keeled; keel visible at the suture on the upper side; aperture broadly lunate, with a sharp and reflected peritreme.

Diameter, ⅝ of an inch; *elevation,* ¼ ditto.

Locality. Red Crag, Walton Naze.

The elegant and, as far as I know, unique specimen, from which the figure above referred to was taken, is from the cabinet of G. S. Gibson, Esq., of Saffron Walden, who has kindly intrusted me with this rarity for publication.   It was found, he informs me, in the interior of a specimen of *Buccinum undatum.*

By careful examination, as well as by the assistance of my conchological friends, I have been unable to find a species with which this could be identified.   In comparing it with *H. rufescens,* the shell to which it appears to have the greatest affinity, some essential differences exist; it is rather larger, and has seven volutions, whereas in that species I have never been able to obtain a specimen with more than six; it is more distinctly and roughly corrugated on the upper surface, and the keel, though obtuse, is visible up the spire; the lines of growth are not more prominent upon the under surface than those upon *H. rufescens,* where there is a strong resemblance between the two.   In this the aperture is of the same form, though the peritreme is rather more reflected, and a little depressed at the suture.

CYCLOSTOMA,* *Lam.* 1799.
CYCLOSTOMA ELEGANS.   *Drap.*

Specimens of this well-known species are found in the upper part of the Coralline Crag, as well as of *Helix crystallina, Pupa marginata,* and *Cyrena trigonula,* in association with marine species.   The locality from which they were obtained is rather a doubtful one, and from their very recent aspect they must, at least for the present, be considered

* Etym. Κύκλος, a circle, and στόμα, a mouth.

as not of the age of the Coralline Crag. The material in which they are found may probably be a disturbed deposit of a more recent period. I have also found this species, with its operculum in position, in the superficial earth resting upon the Red Crag, at the depth of six or eight feet from the surface, but have never seen it in a living state in the eastern part of the county of Suffolk.

SUCCINEA,* <i>Drap.</i> 1805.

AMPHIBULIMA. <i>Lam.</i> 1805.
COCHLOHYDRA. <i>Ferus.</i> 1819.

Gen. Char. Shell ovate, rather elongate; volutions few; spire short; texture thin, fragile, subcorneous, semitransparent; aperture large and entire, longitudinally ovate; peritreme sharp, and confluent with the columella, inner lip spread over a part of the body whorl.

This approaches in general form the genus Limnæa, but may be distinguished by the absence of an oblique fold, which is always more or less distinct upon the columella of the latter. The British species are found about the margins of ponds and ditches, where the surface is moist and damp, and generally upon the stems and leaves of plants. The Succineæ are strictly land animals, although capable of remaining a considerable time under water.

1. SUCCINEA PUTRIS (?) <i>Linn.</i> Tab. I, fig. 5.

HELIX PUTRIS. <i>Linn.</i> Syst. Nat. p. 1249, 1766.
— SUCCINEA. <i>Müll.</i> Verm. 11, p. 97, No. 296, 1774.
— PUTRIS. <i>Mont.</i> Test. Brit. p. 376, t. 16, fig. 4, 1803.
SUCCINEA AMPHIBIA. <i>Drap.</i> Moll. p. 58, t. 3, fig. 22, 23, 1805.
— PUTRIS. <i>Flem.</i> Brit. An. p. 267, 1828.
— AMPHIBIA. <i>Rössm.</i> Icon. p. 91, t. 2, fig. 45, 1835.

S. Testá elongato-ovatá, tenui, fragili; anfractibus tribus convexiusculis, supra depressis, suturis profundis; aperturá ovatá, subverticali.

Shell elongato-ovate, thin and fragile, with about three slightly convex volutions, rather depressed on the upper part; aperture broadly ovate, subvertical; suture moderately deep.

Axis, $\frac{3}{8}$ of an inch.

Locality. Mam. Crag, Bramerton. Recent, Britain.

This species I presume to be rare; but one specimen has come into my possession since the publication of my Catalogue, and this appears to be only a little more elongated than the general form of this species, but in that character it is variable, especially among my specimens from the fluviatile deposit at Stutton, where it is abundant. The artist has given rather too great an expansion to the lower part of the aperture; bringing the resemblance too near to the following species. It is, however, a doubtful identification.

* Etym. Succineus, of amber.

2. SUCCINEA OBLONGA.   *Drap.* Tab. I, fig. 6.

    SUCCINEA OBLONGA.   *Drap.* Moll. p. 59, tab. 3, fig. 24-25, 1805.
    HELIX ELONGATA.   *Daudeb.* Hist. des Moll. pl. 11, fig. 1-3.
    AMPHIBULINA OBLONGA.   *Pfeiff.* p. 68, t. 3, fig. 39, 1821.
    SUCCINEA OBLONGA.   *Rössm.* Icon. p. 92, t. 2, fig. 47, 1835.
    —   *Grateloup.* Cat. des Moll. de Dax, p. 123, 1838.
    —   Nyst Coq. foss. de Belg. p. 466, pl. 39, fig. 11, 1844.
    —   *S. Wood.* Catalogue 1844.

S. Testá elongato-ovatá, tenui ; spirá elevatá, acuminatá ; apice obtuso, suturá profundá ; anfractibus tribus subobliquis ; aperturá ovatá, spiram vix superante.

Shell elongato-ovate, with an elevated and acuminated spire, and an obtuse apex ; volutions three, rather oblique and ventricose, with a deep suture ; aperture ovate, not exceeding the length of the spire.

Axis, ¼ of an inch.

Locality.   Mam. Crag, Bulcham.   Recent, Britain.

One specimen of this species, from the cabinet of Captain Alexander, is all that I have seen.   It appears to correspond with the recent shell, although the suture is not quite so deep.   This species is also found, I believe, abundantly at Maidstone ; some specimens from that locality were given me by Mr. Morris, and I have found the same species, though sparingly, at Clacton, in which specimens the suture is rather deeper than in the Crag shell.

AQUATIC, Flem.

LIMNÆA,* Lam. 1799.

    LIMNEUS.   *Drap.* 1805.
    LYMNÆUS.   *Brand.* 1815.
    LYMNÆA.   *Desmarest.*
    LYMNEA.   *Risso.* 1826.
    LIMNEA.   *Flem.* 1828.
    LYMNUS.   *Montf.* 1810.
    LEPTOLIMNEA.   *Swains.* 1840.
    STAGNICOLA.   *Leach.* 1820.
    GULNARIA.   —   1820.

Gen. Char. Shell ovate or oblong, sometimes elongate and acutely turreted, smooth and naked ; spire always apparent, more or less elevated ; semitransparent and of a subcorneous texture ; volutions convex, somewhat depressed, sometimes tumid and rapidly enlarging ; aperture large, open, and ovate, with a flexuous or plicated columella ; peritreme acute.

Animals of this genus are inhabitants of fresh water, more especially of that which is still and stagnant, although a few have been occasionally found in marshes where

<hr>

    * Etym. Λιμναῖος, of stagnant water.

the water is brackish. The species found in the Crag are, however, such as in the present day are frequenters of ponds and ditches, and were in all probability conveyed to their present locality as dead and empty specimens. Hitherto this genus has not been found in any older formation than the Eocene, during which period it lived and flourished in great profusion.

1. LIMNÆA PALUSTRIS (?) *Linn.* Tab. I, fig. 9, *a—d.*

    HELIX PALUSTRIS. *Gmel.* Syst. p. 3658, 1782.

    LIMNEUS PALUSTRIS. *Drap.* Hist. Nat. des Moll. ter. et fluv. p. 52, tab. 2, fig. 40, 1805.

    HELIX PALUSTRIS. *Mont.* Test. Brit. p. 373, t. 16, fig. 10, 1803.

    LYMNÆA TENUIS. *Woodward.* Geol. of Norf. t. 3, fig. 30, 1833.

    LIMNÆUS PALUSTRIS. *Gray.* 2d edit. Turt. p. 239, fig. 107, 1844.

    LYMNEUS PALUSTRIS. *S. Wood.* Catalogue 1842.

L. Testá elongato-ovatá, tenui: spirá elevatá, apice acuto; anfractibus 4—5 convexis, lævigatis; aperturá ovatá.

Shell elongato-ovate, thin, with an elevated spire and acute apex; whorls about five, convex and smooth, aperture ovate, half the length of shell.

Axis, $\frac{5}{8}$ of an inch.

Locality. Mam. Crag, Bramerton and Bulcham.   Recent, Britain.

Two or three belonging to Mr. Lyell, with one of my own, are all the fossil specimens that I have been able to examine; they do not satisfactorily agree with the above recent species, although they resemble it more than any other with which I am acquainted, and present a sort of intermediate form between it and *L. pereger,* having the elevated spire of *L. palustris* and the convex whorls of *L. pereger,* or perhaps more like gigantic specimens of *L. truncatulus.* The left lip, which in the recent shell is a good deal reflected, is broken away in the Crag specimens, showing a small umbilicus. This species I have found at Stutton and Clacton, precisely resembling the recent form.

2. LIMNÆA PEREGRA. *Müll.* Tab. I, fig. 7, *a—b.*

    BUCCINUM PEREGRUM. *Müll.* Verm. p. 130, No. 324, 1774.

    BULIMUS PEREGER. *Bruguière.* Dict. No. 10, 1789.

    LIMNEUS PEREGER. *Drap.* Moll. pl. 2, fig. 34-37, 1805.

    HELIX PEREGRA. *Mont.* Test. Brit. 1803.

    LIMNÆUS VULGARIS. *Rössmasler.* Icon. p. 97, t. 2, fig. 53-54, 1835.

    — PEREGER. *S. Wood.* Catalogue 1842.

L. Testá ovato-oblongá, tenui, fragili; anfractibus quatuor, convexis; suturis profundis; spirá mediocri, acutá; labio reflexo, aperturá ovatá.

Shell ovato-oblong, thin, and fragile, with four convex volutions and a deep suture; moderately elevated spire, and acute apex; aperture elongato-ovate, with a subreflected peritreme; left lip covering the umbilicus.

Axis, $\frac{1}{2}$ an inch.

Locality. Mam. Crag, Southwold and Bramerton.   Recent, Britain.

One specimen of my own, with a few belonging to Mr. Wigham, obligingly sent to me for the purpose of being figured, are sufficiently characteristic to be fairly identified with one of the recent British forms; and I feel justified in assigning it to *L. peregra*. This is also found at Stutton and Clacton, but I have only met there with the more ovate and less elongated variety.

3. LIMNÆA TRUNCATULA (?)   *Müll.* Tab. I, fig. 8, *a—b*.
        BUCCINUM TRUNCATULUM.   *Müll.* Verm. p. 130, No. 325, 1774.
        HELIX FOSSARIA.   *Mont.* Test. Brit. p. 372, t. 16, fig. 9, 1803.
        LYMNEUS MINUTUS.   *Drap.* Moll. pl. 3, fig. 5-7, 1805.
        —— FOSSARIUS.   *Turt.* Man. fig. 108.
        LIMNEA FOSSARIA.   *Flem.* Brit. An. p. 274, 1828.
        LIMNEUS MINUTUS.   *Rössm.* Icon. p. 100, t. 2, fig. 57, 1835.

    L. Testá pusillá, ovato-conicá, perforatá, tenui, fragili; anfractibus quinque convexis; suturis profundis; spirá elevatá; apice aĉuto; aperturá ovatá.

    Shell small, ovato-conical, perforated, thin, and fragile, with about five convex volutions, and a deep suture; spire elevated; apex acute; aperture ovate; and a slightly reflected pillar lip.

    Axis, ¼ of an inch.

    Locality. Mam. Crag, Bramerton.   Recent, Britain.

    One small specimen, represented at fig. 8 *b*, sent to me by Mr. Wigham for the purpose of description, appears to belong to this species; it has the convex volutions, deep suture, and open umbilicus, corresponding with those of the recent shell.   Fig. 8 *a* is the representation of a specimen from the cabinet of Mr. Lyell of a more doubtful character;[*] although more elongated than the general form of the recent species, it appears to correspond in other respects, and I feel disposed to assign it to *L. truncatulus*.   This species is found in the truly Lacustrine beds of Stutton and Clacton, corresponding precisely with the recent form.

PLANORBIS,[†] Müller, 1781.

    Gen. Char. Shell orbiculato-depressed, discoidal; volutions apparent above and below, convoluted upon a nearly horizontal axis, thin, light, smooth, and somewhat corneous; aperture simple, lunate, crescent-shaped, or subquadrate; outer lip thin, sharp; inner lip slightly spreading over the body whorl.   No operculum.

    The animals belonging to this genus are inhabitants of pure fresh water; their most favoured places of abode being in waters that are stagnant, although a few are found in gentle streams; but none are as yet known in those places to which salt or brackish water has access; the specimens, therefore, found in the Crag must of course have been accidentally introduced, and are few in number.   Shells of this genus are formed upon a nearly horizontal plane, and some considerable disagreement

---

     * Mr. Lyell's specimen measures ⅜ of an inch.
     † Etym. *Planus*, flat, and *orbis*, a circle.

(not yet satisfactorily adjusted) has existed among conchologists as to the correct determination of the upper and under sides of these shells; some have considered them as dextral, while others maintained them to be sinistral; in my descriptions they are considered as dextral, although the artist has made them sinistral.

1. PLANORBIS COMPLANATUS. *Linn.* Tab. I, fig. 10, *a—c.*
    HELIX COMPLANATA. *Linn.* Syst. Nat. p. 1242, 1766.
    PLANORBIS UMBILICATUS. *Müll.* Verm. p. 160, No. 439, 1774.
    HELIX COMPLANATA. *Mont.* Test. Brit. p. 450, t. 25, fig. 4, 1803.
    PLANORBIS COMPLANATUS. *Flem.* Brit. An. p. 278, 1828.
    PLANORBIS MARGINATUS. *Drap.* Moll. p. 45, pl. 2, fig. 11-12, 1805.
    —   *Gray.* 2d edit. Turt. Man. p. 265, fig. 88, 1840.
    —   *S. Wood.* Catalogue 1842.

Pl. Testá discoideá, complanatá, lævigatá: anfractibus supernè convexis, infernè complanatis: carina marginali, non prominenti; spirá depressá.

Shell discoidal, flattish, smooth, with the spire slightly depressed; nearly flat beneath, with the margin carinated but not prominent; volutions convex above; aperture lunate; the upper part of the peritreme projecting beyond the lower.

Diameter, $\frac{3}{8}$ of an inch.

Locality. Red Crag, Butley.
    Mam. Crag, Bulcham.   Recent, Britain.

My cabinet contains but one specimen of this species, found by myself in undisturbed Red Crag. It corresponds with the common recent form of this shell, and has a slightly carinated edge on the flat or lower side, while the volutions on the other side are convex, with the spire depressed and concave, and the edge of the outer lip projecting and curved. Two specimens from the Mam. Crag, near Norwich, were sent to me for description by Mr. Wigham; they do not differ, however, in any respect from my own.

2. PLANORBIS SPIRORBIS. *Müll.* Tab. I, fig. 11, *a—c.*
    PLANORBIS SPIRORBIS. *Müll.* Verm. 11, p. 161, 1774.
    HELIX SPIRORBIS. Gmel. Syst. p. 3624, No. 36, 1788.
    —   *Mont.* Test. Brit. p. 455, t. 25, fig. 2, 1803.
    PLANORBIS VORTEX, var. *β.* *Drap.* Moll. t. 2, fig. 6-7, 1805.
    PLANORBIS SPIRORBIS. *Gray.* 2d edit. Turt. Man. p. 268, fig. 98.
    —   *S. Wood.* Catalogue 1842.

Pl. Testá discoidea, utrinque plano-depressá; anfractibus subcylindraceis, ultimo obsolete angulato.

Shell discoidal, thin, fragile, somewhat concave on both sides, with very slightly increasing, subcylindrical or subquadrate volutions; outer or lower margin obsoletely angulated.

Diameter, $\frac{1}{7}$ of an inch.

Locality. Mam. Crag, Bulcham.   Recent, Britain.

This is figured and described from a single specimen, belonging to Mr. Lyell. Its outer edge is rounded and not carinated, corresponding in that respect with the recent shell, which is now considered by most conchologists to be distinct from *P. vortex*, in which one edge is strongly keeled, and the surface thereby made quite flat, with the suture reduced to a fine and narrow line. Both these species are found in the fresh-water deposit, at Clacton, although *P. vortex* is the most abundant.

3. PLANORBIS CORNEUS. *Linn.* Tab. I, fig. 12, *a—c.*

> HELIX CORNEA. *Linn.* Syst. Nat. p. 1243, 1766.
> — *Mont.* Test. Brit. p. 448, 1803.
> PLANORBIS CORNEUS. *Drap.* Moll. pl. 1, fig. 42, 44, 1805.
> — *Gray.* 2d edit. Turton's Man. fig. 95, 1840.
> — *G. B. Sowerby.* Genera, fig. 1, 18—?
> — *Lam.* 2d edit. Hist. Nat. des An. sans Vert. viii. p. 382, 1837.
> — *Rössmasler.* Icon. t. 2, p. 14, pl. 7, fig. 113, 1838.
> — *Dujardin.* Mém. Soc. Géol. de France, t. 11, 1837.
> — *S. Wood.* Catalogue 1842.

Pl. Testá discoideá, tenui, fragili, plano-depressa, altero laté et profunde umbilicata; anfractibus teretibus rapidè crescentibus; aperturá lunato rotunda.

Shell discoidal, thin, and fragile, rather flat on one side, with a deep or canaliculated suture; umbilicus broad and deep; volutions subcylindrical, rapidly increasing, impressed by the previous whorl.

Diameter, $\frac{3}{8}$ of an inch.

Locality. Mam. Crag, Bulcham. Recent, Britain.

The figure is from a specimen in the cabinet of Capt. Alexander, and the only one I have seen. It corresponds precisely with specimens of the same size, of the common recent British species, and there is no doubt of its identity. The recent shell, in its young state, is marked with fine spiral striæ, and the volutions, in more advanced age, are sometimes angulated. The striæ are more especially distinct in the *perios-traca*, which, when removed, leave the shell nearly smooth. Our specimen has the slightest possible trace of spiral striæ. This species has much the character of a sinistral shell, taking that for the under side in which the spire is most depressed and umbilicated; but the portion of the peritreme is more produced and projecting on that side which, if it were a dextral shell, would be the uppermost. The depression of the vertex appears rather an anomaly in the mode of volution, but it occasionally occurs in species that are convolute upon a horizontal axis, as among some of the *Bullæ*, where the lip is elevated above the spire, giving it there a deep umbilicus. Moreover, I should imagine that the projection of the lower portion of the aperture of the shell would rather impede the action of the foot, and incommode the motions of the animal. I have therefore considered that to be the upper side (fig. 12, *b*), in which the projecting portion of the peritreme would act rather as a protective covering to the animal.

CONOVULUS,* *Lam.* 1812.

MELAMPUS. *Montf.* 1810.
CONOVULA. *Ferus.* 1819.
MARINULA. *King.* 1831.
CONOVULUM. *G. B. Sow.* jun. 1841.
AURICULA. *Lam.*
VOLUTA (spec.) *Linn.*

Gen. Char. Shell oval or elongate, subcylindrical ; with generally a smooth exterior and short conoidal spire ; aperture rather long and narrow ; peritreme continuous, with two or three folds upon the columella ; outer lip sometimes plain, occasionally denticulated within.

This was proposed, in the first instance, by Lamarck, as a division of his extensive genus Auricula, but was afterwards suppressed by himself. Mr. J. E. Gray has recently adopted it in his edition of Turton's 'Manual of Land and Fresh Water Shells,' as a distinction from the Auriculæ, which are land shells. The *Conovuli* live generally in water, and are found sometimes where it is purely fresh ; at others they extend their range into the sea.

1. CONOVULUS PYRAMIDALIS. *J. Sow.* Tab. I, fig. 13, a—b.
    AURICULA PYRAMIDALIS. *J. Sow.* Min. Con. t. 379, 1822.
    —   *Nyst.* Coq. foss. de Belg. p. 473, pl. 39, fig. 12, 1844.
    CONOVULUS PYRAMIDALIS. *S. Wood.* Catalogue 1842.

C. Testá ovato-ventricosá, striatá, crassá ; spirá conoideá pyramidali ; apice obtusiusculo ; anfractibus 7—8 angustis, supernè convexis, subcylindraceis ; aperturá ovato-oblongá auriformi ; columellá biplicatá.

Shell ovato-ventricose, striated? thick and strong, with a conoidal or pyramidal spire, and a slightly obtuse apex ; volutions about seven, elongated and subcylindrical, with an ear-shaped aperture, and two folds upon the columella ; inner lip slightly reflected, generally covering the umbilicus.

Axis, ¾ of an inch.

Locality. Red Crag, Sutton.

    Mam. Crag, Bramerton and Thorpe, near Aldborough.

Found rather sparingly, in the Red Crag, but abundantly in the estuary deposit, at Thorpe, where the specimens are generally more perfect, though seldom so large. It is a thick and strong shell, rather clumsily formed, with a sharp and simple outer lip, somewhat thickened within, particularly at the lower part, and slightly contracted in the middle; the upper part of the volution slightly convex, sloping outwards, the body of the shell nearly straight, contracting towards the base, with two folds upon the inner lip ; the upper one sharp, prominent, and compressed ; nearly at right

---

* Etym. *Conus* and *ovulum.* This name of Melampus has priority, and ought to have been the one employed, but this having been used in my Catalogue is here retained.

angles to the axis of the shell, and situate about the middle of the aperture, at the lower part of the body of the shell; the other upon the columella, midway between the upper fold and the base of the aperture, and is more oblique of the two. The inner lip is thick, especially at the lower part, and reflected, sometimes entirely covering the umbilicus. The specimens are generally quite smooth; but that I imagine is an accidental circumstance, as vestiges of punctured striæ, may be occasionally observed within the aperture, or by the removal of the left lip, and the shell was probably, in its recent state, entirely covered with striæ. There is one deep ridge round the upper part of the volution a little below the suture, and an incipient sinus at the upper angle of the aperture. I am strongly induced to think that the inhabitant was an animal more inclined to salt water than to fresh, as the specimens from the Red Crag have attained to a greater size, and appear to have received their full development as associates with purely marine forms.

2. Conovulus myosotis (?) *Drap.* Tab. I, fig. 14, *a—b. var. β,* fig. 15.

      Auricula myosotis. *Drap.* Hist. des Moll. pl. 3, fig. 16, 17, 1805.
      Conovulus bidentatus. *Gray.* 2d edit. Turt. Man. p. 227, pl. 12, fig. 145, 1840.
      Conovulus myosotis. *S. Wood.* Catalogue 1842.

    C. Testá oblongo-conicá lævigata; spirá elevatá, apice obtusiusculo; anfractibus 8, *convexiusculis; columellá biplicatá; labro tenui, simplici.*

Shell elongated, smooth, with an elevated conical spire, and slightly obtuse apex; whorls moderately convex, with two folds upon the columella; outer lip sharp and plain.

    Axis, $\frac{7}{16}$ of an inch.

    Locality. Red Crag, Sutton.

          Mam. Crag, Bramerton.   Recent, Britain.

Three specimens of this species were found by myself in the Red Crag, associated with deep water, as well as more littoral species, probably carried there by the efflux of the tide upon pieces of decayed wood, its favorite haunt. It has only two folds upon the columella, the upper one is prominent and compressed, placed at nearly right angles to the axis, the lower one is rather more obtuse and oblique. It is larger than any British specimens I have seen, but is less elongated than my recent specimens of *C. myosotis* from France, which have also three folds. Montague says of his *Voluta denticulata* that it is a variable shell, and that specimens have sometimes as many as four folds upon the columella, while others have only two, and that the outer lip is sometimes denticulated within.

The same remark is made by Dr. Gould, who considers it probable that no less than six, perhaps seven, shells, figured and described under different names, are only modifications of this one species, arising from age, accident, and locality.

The Crag shell resembles, in its dentition and ovate form, the recent British species or variety *C. bidentatus,* Gray, but is much larger, having at least twice its length, with a corresponding diameter. In the figure above referred to the artist has scarcely represented it as sufficiently elongated. Fig. 15 is the representation of a shorter and more inflated variety, which might perhaps be considered as a species, with the name of *intermedius*, except that *C. myosotis* has in itself a tendency to extraordinary variation, and coming from a formation in which we are presented in other species with so many monstrous forms, we may be allowed, if necessary, a more than common deviation in this case. The number of teeth upon the columella being three, with a slight difference in their size and inclination, at one time induced a belief that it might be distinct, but the recent species, possessing even a greater degree of variation, will justify this being considered only as a variety. One specimen from the cabinet of Capt. Alexander, and one of my own, are all that I have seen from the Mammaliferous Crag.

Ord. *PECTINIBRANCHIATA*, Cuv.

Sect. *a.* SOLENOSTOMATA, Flem.

Gen. Ovula,* *Bruguière*, 1792.

Amphiceras. *Gronov.* 1781 (fide Gray).
Ovulus. *De Montfort.* 1810.
Simnia. *Leach.* 1819. *Risso.* 1826.
Ovulum. *G. B. Sowerby*, 18—(?)
Bulla (spec.) *Linn.*
Calpurna. *Flem.* 1828.
Birostra. *Swains.* 1840.

Gen. Char. Shell fusiform or oviform, sometimes attenuated or greatly extended at each extremity; externally smooth or very finely striated; convolute upon a nearly horizontal axis, with a hidden spire; aperture elongate and narrow; outer lip generally inflected and denticulated; inner lip smooth.

In form and mode of volution this genus approaches some of the species in the genus Bulla, especially *Bulla acuminata*; but it may be distinguished from that genus by its having both extremities of the aperture canaliculated. In this a fold or tooth-like projection is visible at the upper part above the spire, whereas in Bulla, the fold, when visible, is upon the columella. The surface of the shell, when it has attained its full size, is covered with an enamel-like coat, which is the extension of the columella lip; and it is therefore supposed that the animal has a mantle so extensive as to be able to envelope the entire shell. It differs from Cypræa principally in the ventricose smooth body of the volution, and in the absence of denticulations on the columellar lip.

In a recent state the greater number of known species are natives of a warm climate, and one is quoted as an inhabitant of the British seas. It is not as yet known below the tertiary formations, and two British fossil species have been published, one of which is from the Crag.

---

* Etym. The diminutive of *ovum*, an egg.

1. Ovula Leathesii. *J. Sow.* Tab. II, fig. 1, *a—b.*

     Ovula Leathesii. *J. Sow.* Min. Con. t. 478, 1824.

     —   *Nyst.* Coq. foss. de Belg. p. 605, pl. 43, fig. 19, 1844.

     Calpurna Leathesii. *Flem.* Brit. An. p. 331, 1828.

     Ovula Leathesii. *S. Wood.* Catalogue 1842.

     Ovulum Leathesii. *Morris.* Catalogue, p. 155, 1843.

     Bulla spelta. *Linn.*

    O. Testá elongato-ovatá, fusiformi, medio ventricosá, utrinque subacuminatá; aperturá supernè lineari, infernè sub-effusá; labro intus incrassato, subtus rotundato-angulato; columella supernè uniplicatá.

Shell elongato-ovate, fusiform, smooth, rather ventricose in the middle, slightly acuminated at each extremity; aperture contracted above, and a little expanded at the lower part, with a very short canal; outer lip thickened within, reflected on the exterior; inner lip thin, expanded, with an obtuse fold upon the upper part, above the spire.

    Axis, $\frac{13}{16}$ of an inch.

    Locality. Cor. Crag, Sutton.

               Red Crag, Walton Naze.   Recent, Mediterranean.

This is at present rare in the Red Crag, and I have only a fragment from the Coralline Formation. It is in an altered condition, some of the outer coating has gone, leaving the reflected outer lip separated from the shell. There are some faint traces of denticulations upon the inside of the outer lip, as well as a few striæ upon the lower part of the shell; the inner lip is depressed, slightly concave, with a projecting callosity inward. So far as the altered state of this shell will allow of a fair comparison, I do not see any good specific character by which it can be separated from *Bulla spelta* of Linnæus, which is given as a Touraine fossil by Dujardin (Mém. de la Soc. Géol. de France, 1837, p. 302.)

Gen. Cypræa,* *Linn.* 1740.

    Peribolus (spec.) *Adanson.* 1757.

    Coccinella, *Leach.* 1820.

    Trivia. *Gray.* 1830.

    Gen. Char. Shell oviform, oblong, or subglobular, with a short depressed spire, visible only in the young state; covered with a coating of enamel in the adult. Aperture linear, as long as the shell, crenulated or denticulated on both sides, terminating at both extremities in a short canal, with an inflected and inflated outer lip, generally smooth, sometimes pustulous, or transversely ribbed.

    In my Catalogue the species belonging to this genus were enumerated under the name of Trivia, proposed by Mr. J. E. Gray, for the reception of those species which are transversely ridged or sulcated upon the exterior; as the animals, however, are known not to differ in any essential character from true Cypræa, I have accordingly restored

---

    * Etym. *Cypris,* one of the names of Venus.

them to that genus. The shells of these species are, in the young state, quite smooth, and have a visible spire, with a sharp and simple outer lip, and it is only when full grown that the mantle envelopes the entire shell, depositing a calcareous coating, by which the spire is hidden, the exterior is covered with transverse ridges, and the outer lip is thickened and inverted.

The larger and smoother species are inhabitants of tropical or subtropical regions, but, as a genus, it extends from the equator to the coast of Greenland. A species of the section Trivia has been found in the Eocene formations of this country, one specimen of which is in the cabinet of Mr. Wetherell, and another in that of the late Mr. Channing Pearce.

" Pig" is the common name of these shells upon the coast. In Italy they are called Porcelli ; and Porcelain, the common name of cowries, is taken from the fancied resemblance of these shells to pigs."—(Gray Zool. Proc. 1832.)

1. Cypræa avellana. *J. Sow.* Tab. II, fig. 5, *a—e.*

    Cypræa avellana. *J. Sow.* Min. Con. t. 378, fig. 3, 1832.

    Trivia avellana. *S. Wood.* Catalogue 1842.

    Trivia testudinella, *var. β. S. Wood.* Catalogue.

    Cypræa avellana. *Nyst.* Coq. foss. de Belg. p. 608, pl. 45, fig. 13, 1844.

    Trivia avellana. *Morris.* Cat. of Brit. Foss. p. 164, 1843.

C. *Testá ovato-globosá, crassá, transversim striatá, striis plus minusve numerosis, sulco dorsali interruptis ; aperturá lineari angustatá.*

Shell ovato-globose, variable, thick and strong, covered with transverse striæ or ridges, more or less numerous, interrupted on the back by a longitudinal sulcus ; aperture linear, with the inner lip concave at the lower part.

Axis, $\frac{3}{4}$ of an inch.

Locality. Cor. Crag, Sutton.

            Red Crag, Walton and Sutton.

Rare in the Coralline Crag, but not so in the Red Crag at Walton. It is a very variable species, the length of axis ranging from three fourths to less than one fourth of an inch, with a transverse diameter, in some individuals, equal to its length, while, in others, it does not exceed two thirds of it. The number of ridges upon the exterior varies in almost an equal degree, and affords but little assistance for specific determination. One specimen in my cabinet has as many as forty, while another has only twenty-four ridges. *T. testudinella* of my Catalogue was the elongated var. (fig. 5, *c*), which is now united with *C. avellana,* as I have reason to believe, from the examination of more specimens, that the differences are not specific, and that they may be connected by intermediate forms. It appears distinct from *C. affinis,* Dujard., in which the costæ are more irregular, and terminate obtusely, leaving a naked sulcus on the back, while in this species the ridges are continued across it.

This species is also found in the Touraine beds, specimens of it, from that locality, are in the cabinet of Mr. Lyell.

2. Cypræa affinis. *Dujard.* Tab. II, fig. 9, *a—b.*

Cypræa affinis. *Dujard.* Mém. sur les Conches du Sol en Touraine, Mém. de la Soc. Géol.
de France, 1837, p. 304, pl. 19, fig. 12.
Trivia affinis. *S. Wood.* Catalogue 1842.

*C. Testá ovato-oblongá, transversim striatá; striis sulco dorsali interruptis, et prope
sulcum sub incrassatis; aperturá angustatá.*

Shell ovato-oblong, transversely striated or ridged, interrupted on the back with a
longitudinal furrow; striæ or ridges rather irregular, slightly thickened at their
terminations; aperture elongate.

Axis, ⅜ of an inch.

Locality. Cor. Crag, Sutton.

At present I have only two specimens of this species, from the Crag of Suffolk, and
they appear to differ from *C. avellana* in being of a more elongated form, in the irre-
gularity of the ridges, and in their thickened terminations at the borders of the sulcus.
Mr. Lyell has liberally supplied me with a good series of this species from the Touraine
beds, which, although presenting much variation among themselves, are always less in
size than the British species, and appear to preserve the general character of a naked
sulcus. They may, nevertheless, be only varieties of one species, but I have no specimens that will unite the two, and have in consequence kept them distinct, as in all I
have hitherto seen a difference may be readily pointed out.

3. Cypræa Angliæ. *S. Wood.* Tab. II, fig. 7, *a—b.*

Trivia Angliæ. *S. Wood.* Catalogue 1842.

*C. Testá ovato-globosá, striis, transversis paucis, sulco dorsali interruptis, et prope
sulcum sub incrassatis; apertura lineari subcentrali.*

Shell globosely ovate, with a small number of transverse ridges, interrupted by a
dorsal sulcus; ridges slightly thickened at the edge of sulcus; aperture subcentral;
outer lip thickened and recurved.

Axis, ⅜ of an inch.

Locality. Red Crag, Sutton.

This species is rare. It differs from the two preceding ones in having a less number
of ridges, and in being rather more depressed, with a more orbicular form. The ridges
seldom exceed nine in number, and are carried over the sulcus, but appear a little
thickened at their edges. The sulcus in the figure is scarcely enough displayed.

4. Cypræa retusa. *J. Sow.* Tab. II, fig. 8, *a—b.*

Cypræa retusa. *J. Sow.* Min. Con. t. 378, fig. 2, 1832.
Trivia retusa. *S. Wood.* Catalogue 1842.

*T. Testá ovato-globosá, transversim striatá, striis paucis remotis; sulco dorsali nullo;
aperturá lineatá, submedianá.*

Shell ovato-globose, with few transverse striæ, not exceeding 8—9; spaces between
them broad; no dorsal sulcus; aperture subcentral, nearly straight.

Axis, $\frac{3}{8}$ of an inch.

Locality. Cor. Crag, Sutton.

Red Crag, Sutton.

This is not abundant in either formation. It much resembles the preceding (*C. Angliæ*), in form and paucity of ridges, but there is no longitudinal sulcus on the back. The outer lip is reflected, bringing its linear opening nearly to the centre of the shell. These differences, as well as its more orbicular form, strike the eye immediately, and are, I think, sufficient specific characters. One specimen has only seven ridges across the back.

5. Cypræa Europæa. *Mont*. Tab. II, fig. 6.

    Cypræa pediculus Europæa. *Gmel*. Syst. Nat. p. 3418.

        —   Europæa. *Mont*. Test. Brit. Sup. p. 88, 1808.

        —   Arctica. - - - - p. 200.

        —   bullata, juv. - - - p. 202.

        —   coccinella. *Lam*. An. du Mus. tom. xvi, p. 104, 1810.

        —   — *Nyst*. Coq. foss. de Belg. p. 609, pl. 45, fig. 14.

        —   coccinelloides. *J. Sow*. Min. Conch. t. 378, fig. 1, 1823.

        —   Europea. *Flem*. Brit. An. p. 330, 1828.

        —   Europæa. *Gray*. Zool. Proc. p. 14, 1832.

        —   Norvegica. *Sars*. Beskr. og. Jag. 1835.

    Trivia Europæa. *S. Wood*. Catalogue 1842.

C. Testá ovato-oblongá, ventricosá; sulcis transversis lævibus, sulco dorsali nullo; aperturá regulariter arcuatá, postice subproductá, dentibus æqualibus.

Shell oblong, ovate, ventricose, transversely sulcated, ridges bifurcating, smooth, raised; aperture linear; lower part of inner lip concave, slightly produced.

Axis, $\frac{1}{2}$ of an inch.

Locality. Cor. Crag, Sutton.

Red Crag, *passim*.   Recent, Britain.

This is very abundant in the Red Crag, and exceedingly variable. My largest specimen reaches $\frac{5}{8}$ of an inch in length, while the smallest is little more than $\frac{1}{8}$ inch, with ridges varying on the back from ten to twenty, bifurcating towards the edges. The sulci, as well as the ridges in this and all the preceding species, are quite smooth, the latter being less in width than the former, and generally flat upon the top. Although this species is so abundant, I have never yet been able to procure a young or immature specimen. Mr. Lyell has it also from Touraine.

In describing these shells the term transverse is given to the ridges upon the exterior, while the sulcus upon the back is called longitudinal, corresponding in that character with the position of the animal in its natural state, with the head in a linear direction from the apex of the spire, irrespective of the volutions of the shell.

In England these shells are called "Nuns," and in Scotland "Johnny Groat's Buckie," according to Dr. Fleming. In France they have the name of Poux-de-Mer.

Erato,* *Risso*, 1826.

Voluta (spec.) *Don.*
Cypræa. *Mont.*
Marginella. *De France.*

Gen. Char. Shell ovate, more or less angulated or conoidal, generally smooth, with a short, conical, distinct, but visible spire, and submammillated apex; last whorl large, inflated; aperture linear; outer lip, in the adult state, thickened, inflected, and denticulated within; suture covered with enamel, base slightly emarginate.

Only a few species of this group are as yet known, which have been variously described by different authors either as Volutes, Cyprææ, or Marginellæ, and, as far as the shell may be a guide to its determination, it appears to come between the last two genera, having the inflected lip of the former with the small but visible spire of the latter. Seven recent species have been figured and described by Mr. G. B. Sowerby, jun., in his 'Conchological Illustrations,' and these are all small shells; they have generally a few folds or plicæ at the base of the columella, but are sufficiently characterized to form a distinct and separate genus.

1. Erato lævis. *Don.* Tab. II, fig. 10, *a—b.*

Voluta lævis. *Don.* Brit. Shells, t. 145, 1799.
Cypræa voluta. *Mont.* Test. Brit. t. 6, fig. 7, 1803.
Bulla diaphana, juv. Test. Brit. p. 225, t. 7, fig. 8.
Voluta cypræola? *Broc.* Sub Appen. t. 4, fig. 10, *a—b,* 1814.
Erato cypræola. *Risso.* Hist. Nat. des princip. Prod. de l'Europe, tom. iv, pl. 7, fig. 85, 1826.
Marginella Donovani. *Payr.* Cat. t. 8, fig. 25, 27, 1826.
— cypræola. *Dujard.* Mém. de la Soc. Géol. de France, p. 302, 1837.
Erato lævis. *S. Wood.* Catalogue 1842.

E. Testá subconoideá, vel pyriformi, politá; spirá prominulá, obtusiusculá; anfractibus 3—4 obsoletis; aperturá angustatá, labro extus marginato, intus denticulato; columellá ad basim plicatá.

Shell subconoidal or pear-shaped, smooth and glossy, with a short obtuse spire; volutions obsolete; outer lip thickened without, and toothed within; inner lip slightly denticulated, with two or three folds upon the base of the columella.

Axis, $\frac{3}{8}$ of an inch.

Locality. Cor. Crag, Sutton.

Red Crag, Sutton.   Recent, Britain.

This species is rare, particularly in the Red Crag. It appears to agree in every respect with the recent British shell. The mantle enveloped the entire shell in its adult state, and the sutures of the spire are obliterated by an enamel-like calcareous deposit, which extends over the apex. The outer lip is thickened and contracted in the middle, with a row of very small teeth on the left side; these enlarge into two or three folds at the base of the columella. A ventricose variety of this species is in Mr. Lyell's cabinet, from Touraine.

* The name of one of the Muses (*quæ præsit rebus amatoriis,*) given to this genus probably from its connexion with Cypræa (Cypris).

2. ERATO MAUGERIÆ. *Gray*, Tab. II, fig. 11, *a—b*.

ERATO MAUGERIÆ. *S. Wood*. Catalogue 1842.

— *G. Sowerby*. Conch. Illust. fig. 57, 1841.

E. Testá conoideá, subangulatá, lævigatá politá; spirá brevi, obtusá; anfractibus, tribus; aperturá lineatá, basi subcanaliculato; labro denticulato; columellá subplicatá.

Shell small, conoidal, subangulated, smooth, and glossy; spire short; volutions 3—4 tumid, base subcaniculated; outer lip denticulated, with a few plaits upon the lower part of the columella; outer lip a little thickened in the middle.

Axis, ¼ of an inch nearly.

Locality. Cor. Crag, Sutton.

Red Crag, Sutton. Recent, West Indies.

I have only one specimen from the Red Crag, but it is by no means rare in the Coralline beds. There is, I think, no doubt of its identity with the West Indian shell. It has three folds at the base of the columella, with four or five denticulations above them; the outer lip is rather thickened in the middle. One of my specimens from the Cor. Crag has a tinge of colour remaining in the outer lip. The principal differences between this and the preceding (*E. lævis*) are its size and its more tumid and angular form of volution. The figures are slightly enlarged. This species is also from Touraine, in Mr. Lyell's cabinet.

VOLUTA,* *Linnæus*, 1767.

MITRA. *Flem*. 1828.

FASCIOLARIA. *Conrad*.

HARPULA. *Swains*. 1840.

Gen. Char. Shell subovate, or elongato-fusiform, more or less ventricose, sometimes angulated, thick, strong, and generally large, smooth, striated, or tuberculated; apex obtuse or mammillated; outer lip simple, sometimes thickened within; aperture generally large and linear, terminating in a short and deep notch; columella with several folds, of which the lowest is the largest.

Shells of a truly mammillated apex are not as yet known below the Miocene formations. Those species hitherto considered as Volutes, from the Eocene deposits, have an elevated spire and an acuminated apex; they differed probably in their animal inhabitants, and may constitute another genus.

A species in Mr. Lyell's cabinet from the Eocene formation, at Claibourne, Alabama, and also one in the possession of Mr. Edwards, from Bracklesham, appear to preserve an intermediate character, having a small papilliform apex, showing the transition from the acute to the obtuse.

The recent species belonging to this genus are generally natives of the warmer regions of the globe, although one large species of true Volute, with an emarginate base, has been found on the southern coast of Patagonia, in lat. 51° S.

* Etym. *Volutus*, rolled, *a volvo*.

1. Voluta Lamberti.   *J. Sow.* Tab. II, fig. 3, *a—b.*

          —   *Dale.* Hist. of Harwich, pl. 10, fig. 14, 1730.

Volute of Harwich.   *Park.* Organ. Rem. t. 5, fig. 13, 1811.

Voluta Lamberti.   *J. Sow.* Min. Conch. t. 129, 1816.

Mitra Lamberti.   *Flem.* Brit. An. p. 333, 1828.

Fasciolaria Lamberti.   *Conrad.* Journ. Nat. Sc. vol. vi, p. 216.

Voluta Lamberti.   *Charlesworth.* Mag. Nat. Hist. p. 37, fig. 7, 1837.

          —   *Grateloup.* Cat. des An. de l'Adour, p. 50, 1838.

          —   *Dujard.* Mém. de la Soc. Géol. de France, tom. ii, pt. 2, p. 300, 1837.

          —   *Nyst.* Coq. foss. de Belge, p. 587, pl. 45, fig. 4, 1844.

          —   *Morris.* Catalog. of Brit. Foss. p. 167, 1843.

          —   *S. Wood.* Catalogue 1842.

V. Testá fusiformi, ovatá, vel elongatá, tenuissime striatá, spirá conicá, apice papillari ; anfractibus convexiusculis, superne depressis; aperturá ovatá, ad basim subcanaliculatá, attenuatá, vix emarginatá, columellá rectá, plicatá.

Shell fusiform, ovate, or elongated, very finely striated, with 5—6 slightly convex volutions ; aperture ovate ; base subcanaliculated, rather produced ; columella straight, with four plaits.

Axis, 7 inches.

Locality.   Cor. Crag, Aldborough, Ramsholt.

          Red Crag, *passim.*

This species is not confined to any locality, but is found in most places where a section of the Crag is visible. It scarcely fulfils the conditions required by the above generic characters, the base being produced and not emarginate, and the lower plait upon the columella being the smallest. As similar characters may be observed in some recent species, still included in this genus, it may be as well to leave the Crag one where it has been so long placed, until further information be obtained respecting their inhabitants.

It presents a good deal of variation in the proportionate dimensions, some specimens being much elongated, with a longitudinal diameter three times that of its transverse, while in others it is only as long again. The outer lip, when perfect, is sharp and rather arched, with an incipient sinus at the suture. The shell was probably covered with fine striæ, in its original state, but very little of those markings can ever be seen; they are most visible on those specimens that are washed up on the beach at Felixstow. This important character was first pointed out by Mr. Charlesworth, at the above reference.

The length of my largest specimen is seven inches; another imperfect one indicates a magnitude of at least two inches more. Some fine specimens of a Volute, from Touraine, in Mr. Lyell's cabinet, belong, I think, to this species; they are rather thicker than the Crag ones, and short, resembling var. *a* (of our plate), and the folds upon the columella are rather less oblique, but not sufficiently different to remove it from this species. A young specimen shows the fine transverse striæ.

MITRA, *Lam.* 1801.

Gen. Char. Shell turreted, elongated; spire elevated, apex generally acute; volutions smooth, striated, sulcated, or, more frequently, costated; columella slightly recurved and plicated; folds less oblique than in Voluta, the lower fold always the smallest; aperture narrow and elongated; outer lip slightly thickened, and often dentated.

The greater part of the shells constituting this extensive genus, in the recent state, are natives of tropical or subtropical regions, but its geographical range is very extended, one species inhabiting the Greenland seas. Fossil species are by no means abundant, although they are occasionally present in all the tertiary deposits. The range, in depth, of many of the species is very considerable. Professor E. Forbes dredged up *Mitra ebena* in the Mediterranean, at depths varying from twenty to eighty fathoms, and some are even littoral species.

1. MITRA PLICIFERA. *S. Wood.*

MITRA PLICIFERA. *S. Wood.* Catalogue 1842.

The shell that stands under this name in my Catalogue is a true Mitra, possessing four folds upon the columella, the upper one the largest and most isolated, while the lower one is small and nearly obsolete. It is obtusely costated upon the volutions, and transversely striated. My specimens are unfortunately in bad condition, unfit for figuring, and all further notice respecting it must be deferred until some better preserved individuals are obtained.

RINGICULA,* *Deshayes,* 1838.

AURICULINA. *Grateloup.* 1838.
MARGINELLA. *Ménard de la Groye.*
PEPIDES. *Dujardin.*
AURICULA (spec.) *Lam.*

Gen. Char. Shell small and ovate, with a short spire and acute apex; smooth or striated externally; columella callous, deeply plicated; outer lip thickened and reflected, with a deep notch at the base of the aperture.

This is a marine genus, and on that account separated by Deshayes from Auricula, in which these shells had been previously placed by Lamarck. It is in all probability one of the zoophagous molluscs, as the deep notch for the syphon would seem to indicate, though its place in a natural arrangement has not been well determined. Few species are as yet known, and those sparingly distributed throughout the tertiaries; one is still living in the Mediterranean and one at the Gallapagos islands.

* Etym. The diminutive of *ringens* (from *ringo, ringere,* to grin), the specific name of the type of the genus. The name of *Auriculina* was proposed by Grateloup, in 1838, for this genus, but the above name had been previously (?) used by Deshayes, in the same year, in the 2d edition of Lam. Hist. des An. sans Vert., vol. viii.

1. RINGICULA BUCCINEA. *J. Sow.* Tab. IV, fig. 2, *a—b.*

     VOLUTA BUCCINEA.   *Broc.* Sub Appen. pl. 4, fig. 9, 1814.

     AURICULA BUCCINEA.   *J. Sow.* Min. Conch. t. 465, fig. 2, 1823.

     MARGINELLA AURICULATA.   *Dubois de Montp.* pl. 1, fig. 15, 16, 1831.

     PEDIPES BUCCINEA.   *Dujard.* Mém. Soc. Géol. de France, tom. ii, pt. 2, p. 277, 1837.

     AURICULINA RINGENS, *var. β.*   *Grat.* Mém. sur la Fam. des Melan. p. 12, 1838.

     RINGICULA BUCCINEA.   *Nyst.* Coq. foss. de Belge, p. 604, pl. 45, fig. 12, 1844.

     —   *S. Wood.* Catalogue 1842.

R. Testá subovatá, lævigatá, apice acuto, spirá elevatá; anfractibus, 5—6, subinflatis; basi emarginatá; columella plicis tribus acutis; labro marginato, calloso, in medio inflato, non crenato.

Shell subovate, smooth, with an acute and rather elevated spire; volutions, 5—6, very slightly inflated; columella with three folds; base short, emarginate; outer lip thickened; callous in the middle, not crenated; inner lip spreading on the body whorl.

Axis, ¼ of an inch.

Locality. Cor. Crag, Sutton.

         Red Crag, Sutton.

There is, in fact, but one acute and prominent tooth upon the columella, the upper one is a thickening or compression of the inner lip upon the body of the shell, visible only within the mouth; the lower one is the replication of the lower part of the columella. This is very abundant at the first-mentioned locality, where there is also a small variety not more than one eighth of an inch in length.

2. RINGICULA VENTRICOSA. *J. Sow.* Tab. IV, fig. 1, *a—b.*

     AURICULA VENTRICOSA.   *J. Sow.* Min. Con. t. 465, fig. 1, 1823.

     AURICULINA RINGENS.   *Grat.* Foss. de Bord. p. 11, pl. 6, fig. 6, 7, 1838.

R. Testá ovatá, inflatá, tumidá; transversim sulcatá, spira brevi, acutá; columellá triplicatá; labro expanso, adnato, calloso, non crenato.

Shell ovate, tumid, inflated, sulcated along the volutions; spire short; apex acute; outer lip thickened, without crenulations; inner lip expanded.

Axis, ⅜ of an inch.

Locality. Cor. Crag, Sutton.

         Red Crag, Sutton.

Abundant in the Red Crag, and very scarce in the Coralline. This and the preceding shell are considered as varieties by some conchologists. I am induced to keep them separate for the following reasons. This is shorter, broader, more inflated in the volutions than *R. buccinea;* the exterior is deeply sulcated, and the two uppermost folds are placed further apart; the two lower plaits not reaching half-way up the mouth, whereas, in *R. buccinea,* the three plaits are nearly equidistant, and the two lower ones occupy at least two thirds of the mouth; the form of the outer lip in this shell is also more quadrate, and the inside of the outer lip is not so much thickened, nor so callous in the centre. Among all my numerous specimens there are no intermediate forms by which they can be fairly united.

COLUMBELLA,* *Lam.* 1799.

BUCCINUM (spec.) *Adans.*
PYGMÆA. *Humph.* 1797.
COLUMBUS. *Montf.* 1810.
CONIDEA. *Swains.* 1840.

Gen. Char. Shell generally thick, strong, ovate, or somewhat angular, with a short spire, though sometimes turriculate and elevated ; aperture oblong ; base emarginate ; outer lip thickened and dentated within ; generally more or less tumid in the middle ; inner lip irregularly crenulated, thickened posteriorly, and callous ; operculum small and corneous.

Shells of this genus have a tropical or subtropical character, and may be considered almost as a recent group ; a few species have, however, been found in the Touraine beds. The Crag one is rather of an aberrant form, connecting it with Nassa.

1. COLUMBELLA SULCATA. *J. Sow.* Tab. II, fig. 2, *a—d.*

BUCCINUM SULCATUM. *J. Sow.* Min. Con. t. 375, fig. 2, 1823.
— *var. β.* Min. Con. t. 477, fig. 4.
COLUMBELLA SULCATA. *S. Wood.* Catalogue 1842.

C. Testá turritá, elongatá ; spirá elevatá, apice acuto, anfractibus numerosis, convexis, transversim sulcatis ; aperturá elongatá ; labro incrassato, intus denticulato ; basi truncatá, emarginatá ; labro infernè plicato.

Shell elongate, turreted, and sulcated ; spire elevated, apex acute, whorls 8—10, convex ; aperture elongate, subtrapezoidal ; outer lip nearly straight, thickened, and denticulated within ; base short, open, and truncated ; left lip thin, crenulated.

Axis, 1¼ inch.

Locality. Red Crag, Walton Naze. Var. β, Sutton.

Abundant at Walton, but rare in all other parts. I am not aware of its having been found in the Coralline Crag. It is very variable in its proportionate dimensions. In one specimen the mouth is nearly one half its length, while in another it is not more than one fourth. The outer lip is thickened and denticulated within, when in its adult state, and no doubt the left lip was also crenated when recent. One specimen shows a few folds upon the lower part of the columella, with a slight thickening in the middle of the outer lip. These characters are considered sufficient to retain it in the above genus.

ROSTELLARIA,* *Lam.* 1801.

STROMBUS (spec.) *Linn.*

Gen. Char. Shell turreted or fusiform, with numerous volutions, generally ornamented on the exterior by longitudinal costæ or transverse striæ ; aperture ovate, with

* Etym. (?) The diminutive of *Columba,* a dove.
† Etym. *Rostellum,* a little beak.

its superior extremity extended in the form of an elongated and very narrow canal; base with sometimes a lengthened siphonal canal, and a sinus in the lower part of the outer lip.

ROSTELLARIA PLURIMACOSTA.   *S. Wood.* Catalogue of Crag Shells, 1842.

A few worn specimens of a species of Rostellaria are in my cabinet, from the Red Crag of Sutton. They are in a mutilated condition, and unfit for fair comparison; they bear a resemblance, and may possibly be the *R. lucida*, J. Sow. (Min. Con. t. 7), perhaps washed out of the London clay.

APORRHAIS,* *Aldrovandus.*
ROSTELLARIA. *Lam.* 1801.
CHENOPUS. *Phil.* 1836.

Gen. Char.   Shell turreted or fusiform, thick and strong, generally ribbed, nodulous, or carinated; aperture ovate or elongate, terminating in a canal with a calcareous pointed process at the base, and having, in the adult state, an expanded, angularly lobed, or digitated outer lip, sinuated at the lower part near the canal. Operculum corneous.

The name for this genus was used by Aldrovandus in 1623, and by Petiver in 1702, when describing the *pes-pelicani* with the following words: "Aporrhais Edinburgensis minor nodosa." (Gazophylacium, fol. 17, Tab. 79, f. 6, and Tab. 127, f. 11, Cat. No. 85.)   Da Costa (History of British Shells, 4to, 1778) characterizes the same species under the name of Aporrhais with a new specific appellation, and a reference to Petiver.   In the 'Philosophical Transactions' for 1823, Dillwyn uses the same name as a generic term, and applies it to the *pes-pelicani;* and this may be considered as a fair revival of a name proposed before the time of Linnæus.   This species, upon which the genus has been established, was included by Lamarck in Rostellaria, in consequence of the sinus in the lower part of the outer lip resembling the shells of that genus.

In 1836, M. Philippi (Enum. Moll. Sic. vol. i, p. 215,) described the animal of *pes-pelicani*, which, he says, is decidedly different from that of Rostellaria, and proposes, in consequence, to erect it into a genus, under the new name Chenopus, with the *pes-pelicani* as its type.

It has been contended that the genus was in the first instance established upon the form of the shell alone, without a knowledge of its animal inhabitant; and that a detection of a difference in some of the soft parts, with a publication of the anatomical details, will justify the rejection of an old established name and the substitution of a new one.   A principle which, if admitted, will endanger the stability of many other genera that have been formed upon the shell alone, and priority of date, as now considered, will be no security for an author's name.

---

* Probably derived from ἀπόῤῥαξ or ἀποῤῥωξ, *rent, torn*, in allusion to the ragged or digitiform processes of the outer lip.

1. APORRHAIS PES-PELICANI. *Linn.* Tab. II, fig. 4, *a—b.*
 STROMBUS PES-PELICANI. *Linn.* Syst. Nat. p. 1207, 1767.
 TRITONIUM PES-PELICANI. *Müll.* Zool. Dan. prod. p. 244, 1776.
 APORRHAIS QUADRIFIDUS. *Da Costa.* Brit. Conch. p. 136, t. 7, fig. 7, 1778.
 ROSTELLARIA PES-PELICANI. *J. Sow.* Min. Conch. t. 558, 1827.
  — *Dubois de Montp.* Conch. foss. du Plat. Volhyn. Podol. pl. 1, fig. 31, 1831.
  — *Nyst.* Coq. foss. de Belge, p. 561, pl. 43, fig. 7, 1844.
 CHENOPUS PES-PELICANI. *Phil.* En. Moll. Sic. p. 215, 1836.
 APORRHAIS PES-PELICANI. *S. Wood.* Catalogue 1842.
  — *Morris.* Cat. of Brit. Foss. p. 138, 1843.

A. Testá turritá; anfractibus convexis, subangulatis, medio nodulosis, striatis; labro dilatato, in tres digitos partito; digitis divaricatis, canali, elongato, obliquo.

Shell turreted, with 10—11 convex volutions, covered in the centre with small nodules; transversely striated; outer lip digitated, and separated into three sharp digitiform processes, one of which runs up the spire; base of volution angulated; canal produced and curved.

Axis, $\frac{1}{2}$ inch.

Locality. Cor. Crag, Ramsholt and Gedgrave.

Red Crag, Sutton, Newbourn, Brightwell, and Bawdsey. Recent, British Seas.

This shell is rarely found in good condition in the Red Crag; some well-preserved specimens, however, show a perfect identity with the long known recent species. The middle of the volution is slightly angular and nodulous, so also is the prominent keel at the base of the whorl, like that of the recent shell, with another ridge beneath it. The striæ are obliterated from all my Red Crag specimens, though shown in those from the Coralline. This is a living Mediterranean species, and stated by Philippi to be exceedingly variable there. My Crag specimens are very uniform in character.

TEREBRA,* Adanson, 1757.

 TEREBRA. *Lam.*
 SUBULA. *Blainville.*

Gen. Char. Shell subulate, turriculate, generally much elongated, with an acuminated apex; volutions numerous; aperture comparatively short, ovate; base emarginate, with or without a canal; outer lip thin; columella often oblique and spiral, sometimes striated; many species have a corneous operculum.

This as a recent genus is abundant in species, and is nearly confined to the tropical seas. Mr. J. E. Gray has separated several recent species included by Lamarck in this genus, in consequence of a difference in the animal, and united them into a distinct group, under the name of Bullia. There are no good generic characters in the shell of Bullia by which it can be distinguished, although its general form is intermediate between the elongated Terebræ and some of the Nassæ.

* Etym. *Terebro,* to bore.

1. TEREBRA INVERSA. *Nyst.* Tab. IV, fig. 3, *a—b*.

       TEREBRA INVERSA.   *Nyst.* Coq. foss. de Belge, p. 581, pl. 44, fig. 9, 1844.

       —   HETEROSTROPHA.   *S. Wood.* Catalogue 1842.

T. Testá turritá, subulatá, sinistrorsá; longitudinaliter plicatá, anfractibus sub-planulatis, prope suturam appressis, inferne convexiusculis; aperturá ovatá sub effusá; columellá contortá; labro acuto, intus nitido.

     Shell fusiform, turriculate, sinistral, with a rather obtuse apex; whorls nearly flat, longitudinally plicated; aperture ovate; outer lip sharp, with a slightly recurved canal.

     Axis, nearly ¾ of an inch.

     Locality.   Cor. Crag, Gedgrave.

                 Red Crag, Sutton.

     In a communication from M. Nyst, he says that the Belgian shell, which I have not seen, is perfectly identical with that from the Crag. I have therefore given his name; mine, being without description, can be considered only as provisional. My specimens are few, and not in good condition; one of the most perfect shows a ridge running round the upper part of the volution, like that in *T. pertusa*, but not so distinct; neither are the volutions so flat as in that shell, and it appears to have a longer canal.

2. TEREBRA CANALIS.   Tab. IV, fig. 4.

       TEREBRA CANALIS.   *S. Wood.* Catalogue.

     I have only three or four imperfect specimens of what stands under this name in my Catalogue. It much resembles the preceding in the length of its canal and form of its volutions, and may possibly be only a dextral variety; but better specimens than I possess are necessary for such a determination.

     The figure above referred to is a restored form by the union of two imperfect specimens.

CASSIDARIA,* Lam. 1812.

MORIO.   Montf. 1810.

ECHINORA.   Schum. 1817.

Gen. Char. Shell ovate, ventricose, with a short turreted spire, last whorl much inflated, for the most part spirally grooved or tuberculated; aperture large, terminating anteriorly in a slightly recurved canal; outer lip thickened, reflected, and dentated within, in the adult state; inner lip broad and expanded over the umbilicus, with part of its lower edge free.   Operculum (?)

     This genus in most of its characters resembles Cassis, but is distinguished by its rather lengthened and less suddenly recurved canal. In Cassis it is emarginate, short, and reflected. The Crag shell was first placed in the latter genus in the 'Mineral Conchology,' the canal not having been seen by the author of that work.

     This genus is rather indicative of tropical or subtropical regions, and several species have been described from the Eocene as well as from the Meiocene formations of Europe.

---

     * Etym. *Cassida*, a helmet. The name of *Morio* appears to have priority of date. *Cassidaria* having been used in my Catalogue, it is not thought necessary to alter it.

1. CASSIDARIA BICATENATA. *J. Sow.* Tab. IV, fig. 5, *a—b.*

    CASSIS BICATENATUS. *J. Sow.* Min. Con. t. 151, 1824.

    CASSIDARIA BICATENATA. *S. Wood.* Cat. of Crag Shells in Ann. and Mag. Nat. Hist. p. 538, 1842.

    —   *Nyst.* Coq. foss. de Belge, p. 565, pl. 46, fig. 6, 1844.

C. Testá ovato-ventricosá, transversim sulcatá; sulcis depressis, longitudinaliter decussatis; anfractibus convexis, subcarinatis; carinis tuberculosis; labro incrassato, intus obscurè dentato.

Shell ventricose, ovate, sulcated or ridged; obscurely decussated; volutions tumid; last whorl large, inflated, upper part subcarinated and tuberculated; outer lip thickened, faintly dentated; inner lip broad and expanded; columella finely and irregularly striated.

Axis, 4 inches; *transverse diameter,* nearly 3 inches.

Locality. Cor. Crag, Ramsholt and Gedgrave.

    Red Crag, Sutton, Bawdsey, and Felixstow.

This handsome shell is at present exceedingly rare in both formations. Specimens have been occasionally washed up on the beach at Felixstow, along with *Voluta Lamberti,* apparently thicker and stronger than the generality of shells from the Red Crag, which Mr. Charlesworth (Mag. Nat. Hist., 1837, p. 38) thinks were from a deposit posterior in age to that formation.

The subangulated appearance on the upper part of the volution is produced by a double row of nodules upon the ridges, the upper one of which is the most prominent. These nodules may be traced upon two or three of the ridges, becoming more obscure as they descend upon the body whorl. The outer lip of one specimen is much thickened and a large callus deposited upon the body of the shell, forming a distinct sinus at the upper part of the aperture. The canal is rather shorter than in the generality of shells in this genus, but is not deep and reflected like that in Cassis. This shell thickens its outer lip at irregular periods of growth, leaving thereby obtuse varices visible upon the spire. This is also the case in many of the Helmets. Specimens from the English Crag attain a magnitude considerably beyond that figured by M. Nyst; one of mine exceeds it twofold in linear dimensions.

NASSA,* *Lam.* 1801.

BUCCINUM (spec.) *Linn.*

Gen. Char. Shell ovate, turriculate or conoidal; spire generally longer than the aperture; apex mostly acute; exterior grooved, striated, granulated, or tuberculated, sometimes smooth; inner lip largely expanded, with a tooth-like projection at the base of the columella, and sometimes a tooth upon the left lip, at the upper part of the

* *Nassa* is the name of a kind of net, for catching fish, made of twigs like an eel-pot; probably in allusion to the generally reticulated surface of these shells.

aperture ; outer lip generally thickened and dentated within, with a short and some-what reflected canal ; operculum corneous.

This genus is not admitted by all conchologists as a necessary separation from the Linnæan genus Buccinum, and in fact Lamarck subsequently reunited them without assigning any reason for so doing. The principal distinctive character given, is the prominent tooth at the base of the columella, and, it might also be added, the denticulations within the outer margin of the aperture, and the expanded form of the inner lip. There certainly is a close resemblance between some species of each, but such is the case with all approximate genera in their aberrant forms. Two or three species have been figured from the Green-sand as belonging to this genus, and a few doubtful shells are given from the Eocene formations. The Crag deposit is exceedingly rich not only in individual specimens, but also in number of species. The range of this genus in depth seldoms exceeds sixteen fathoms, while some of the species are found under stones at low water.

1. Nassa labiosa. *J. Sow.* Tab. III, fig. 8, and Tab. VII, fig. 22.

     Buccinum labiosum. *J. Sow.* Min. Con. t. 477, 1824.

     —   *Nyst.* Coq. foss. de Belge, p. 577, pl. 43, fig. 14, 1844.

     Nassa labiosa. *S. Wood.* Catalogue 1842.

N. Testá ovatá ; spirá elevatá ; transversim profundè sulcatá ; apice acuminato ; anfractibus septem convexiusculis ; suturis profundis, canaliculatis ; labio extenso ; labro intus denticulato ; canali brevi.

Shell ovate, with an elevated and acuminated spire, ridged or sulcated along the volutions, with a small canal at the suture ; inner lip thick and extended on the body whorl ; outer lip denticulated within ; canal short.

Axis, ¾ of an inch.

Locality. Cor. Crag, Gedgrave.

           Red Crag, Sutton and Newbourn.

Not very abundant. When perfect the spiral ridges are generally regular, flat, as broad as, and sometimes broader than, the spaces between them. The inner lip is much spread and smooth, and the upper angle of the mouth slightly elevated, producing thereby a small but distinct canal at the suture ; canal short, slightly recurved, with about a dozen denticulations within the outer lip. It has a fold or ridge at the base of the columella, with two or three wrinkles on the lower part of the left lip. This is given by Professor E. Forbes, in his 'Report upon the Geological Relations of the existing Fauna and Flora of the British Isles,' as a synonym to *B. semistriatum,* Broc. If it be the same species, it is very differently ornamented upon the exterior. The Crag shell is more elongated than any of my specimens of *B. semistriatum,* and it is regularly striated or ridged all over, and rather more strongly so upon the body of the shell. In *B. semistriatum* it is strongly and coarsely striated only at the base, nor are there any longitudinal folds upon the upper volutions ; lines of growth are faintly visible over the shell.

In Table III, fig. 8, the artist has given too great an expansion to the lower portion of the outer lip. Table VII, fig. 22, is a more correct representation.

2. NASSA INCRASSATA. *Müll.* Tab. III, fig. 4.

TRITONIUM INCRASSATUM. *Müll.* Prod. No. 2946, 1773.
BUCCINUM MACULA. *Mont.* Test. Brit. p. 241, t. 8, fig. 4, 1803.
— ASCANIAS. *Brug.* ex Phil. En. Moll. Sic. vol. ii, p. 188.
— ASPERULUM. *Broc.* t. 5, fig. 8, 1814.
NASA INCRASSATA. *Flem.* Brit. An. p. 340, 1828.
NASSA INCRASSATA. *S. Wood.* Catalogue 1842.

N. Testá ovato-conicá, subturritá; spirá acutá; longitudinaliter costatá; transversim angustè striatá; anfractibus 6—7 convexis, tumidis; aperturá subrotundá; labro incrassato, varicoso, intus denticulato.

Shell rather small, ovato-conical, subturreted, longitudinally costated, and finely striated transversely; whorls 6—7 convex, tumid; aperture subcircular, with a thickened outer lip, denticulated within.

Axis, $\frac{1}{2}$ of an inch, nearly.

Locality. Cor. Crag, Sutton.

Red Crag, Sutton. Recent, Britain.

Very scarce in the lower formation; more abundant in the Red Crag. My specimens seldom exceed $\frac{3}{8}$ of an inch in length. The costæ in the last volution vary from 12—15, with 12—15 threads along the whorls, and carried over the ribs. Denticulations within the lip, lower one the most produced, with a distinct ridge on the opposite side of the canal at the base of the columella, contracting the entrance to the canal, with a prominent tooth upon the body whorl, near the upper part of the aperture. This appears to correspond with the recent species in every character except its size, the recent British specimens being generally larger.

3. NASSA GRANULATA. *J. Sow.* Tab. III, fig. 3.

BUCCINUM GRANULATUM. *J. Sow.* Min. Con. t. 110, fig. 4, 1815.
— *Nyst.* Coq. foss. de Belge, p. 575, pl. 43, fig. 11, 1844.
NASSA GRANULATA. *S. Wood.* Catalogue 1842.

N. Testá ovato-conicá, turritá, apice acuto; anfractibus convexis, longitudinaliter costatis, transversim striatis; striis elevatis gránulatis; aperturá subrotundá; labro incrassato intus dentato; labio supernè unidentato.

Shell ovato-conical, with an elevated and acute spire; whorls convex, longitudinally costated, and decussated by coarse and elevated striæ; aperture subcircular, with a reflected left lip and thickened margin denticulated within.

Axis, $\frac{5}{8}$ of an inch.

Locality. Cor. Crag, Sutton.

Red Crag, *passim.*

Very abundant in the Red Crag, but rare in the Coralline. In shape it much resembles the preceding species, but the striæ, or rather ridges, are fewer and larger, and carried over the costæ, which give it a granulate appearance, and regularly decussate the exterior. The outer lip has fewer teeth. The tooth on the body whorl forms an imperfect sinus at the upper angle of the aperture. *Buc. graniferum*, Dujardin (Geol. Trans. of France, 1837, vol. ii, part 2, pl. 20, f. 11, 12), strongly resembles this species, but appears to have the costæ more apart; in this shell they are contiguous, with a difference also in the striæ. *Buc. granulatum*, Philippi (En. Moll. Sic. vol. i, p. 226, pl. 11, f. 22), appears a different species. The shell figured (fig. 3) is rather above the ordinary size, which rarely exceeds half an inch.

4. Nassa propinqua. *J. Sow.* Tab. III, fig. 2.

> Buccinum propinquum. *J. Sow.* Min. Con. t. 477, fig. 2, 1824.
> — costulatum. *Broc.* t. 5, fig. 9, 1814.
> Nassa propinqua. *S. Wood.* Catalogue 1842.

N. Testá elongato-ovatá, longitudinaliter costulatá; costis (circa 23) nodulosis, striis transversis decussatis; anfractibus convexis; striis prope suturam profundioribus exaratis, angulatis; labio supernè unidentato; labro crasso intus denticulato.

Shell elongato-ovate, longitudinally costated, with about twenty-three small costæ, sulcated or ridged; ridges carried over the ribs, and nodulous; one ridge isolated and distinct near the suture; one tooth upon the body whorl within the mouth; outer lip thickened and denticulated within.

Axis, ¾ of an inch.

Locality. Red Crag, Sutton and Walton Naze.

This species is by no means rare, though seldom found in good condition. Its distinguishing character is the granulated ridge near the suture, which is separated by a wider sulcus than any of the others. The outer lip is thickened only in its adult state; the shell is, however, sometimes ornamented with an obtuse varix, where the animal has enlarged its shell, and has produced another half volution. It is possible this may be the *Buc. costulatum*, Brocchi, which I do not know; the figure given by that author is rather more elongate than our shell.

5. Nassa elegans. *Leathes.* Tab. III, fig. 1.

> Buccinum elegans. *J. Sow.* Min. Con. t. 477, fig. 1, bene, 1824.
> — *Nyst.* Coq. foss. de Belge, p. 576, pl. 43, fig. 13, non bene.
> — *Morris.* Catalogue of Brit. Foss. p. 139, 1843.
> Nassa elegans. *S. Wood.* Catalogue 1842.

N. Testá turritá, angustá, longitudinaliter costatá, transversim striatá; striis paucis elevatis; apice acuminato; anfractibus 7—8 convexis, suturis profundis; aperturá orbiculari; labio supernè uniplicatá, labro intus denticulato.

Shell turreted and elongate, with an elevated spire, costated longitudinally, 10—11 in the last volution, striated or ridged transversely; volutions convex, and suture deep; outer lip thickened, dentated within.

Axis, ¾ of an inch.

Locality. Red Crag, Walton Naze.

I have not seen this shell from any other locality, although in the Cliff at Walton it is by no means rare. The costæ are about 11 in number on the last volution, and stand apart; they are not so wide as the spaces between them, and are crossed by about ten sharp and elevated striæ, which are carried over the ribs, and the mouth has a thickened varix when full grown. This species appears subject to very little variation; among all my specimens there is but a trifling difference. It is more distinctly costated and more elongated than the two preceding. The costæ in the figure given by M. Nyst are represented as close together. In the English specimens there is a considerable space between them. The mouth is subcircular, with about a dozen denticulations within the outer lip, and sometimes three or four at the lower part of the columella. In one specimen the whole of the left lip is covered with denticulations.

6. NASSA CONSOCIATA. *S. Wood.* Tab. III, fig. 7.

NASSA COSTULA. *S. Wood.* Catalogue 1842.

N. Testá turritá, angustá, elegante; spirá elevatá, apice obtusiusculo, longitudinaliter costatá, costis 9—10 rectis, transversim crenulatis; aperturá orbiculari; labio supernè uniplicato; labro incrassato intus rare denticulato.

Shell turreted, with an elevated spire, and rather obtuse apex, longitudinally costated, costæ 9—10 transversely crenulated; aperture orbicular, with a tooth on the left lip; base rather contracted and reflected, outer lip dentated within.

Axis, $\frac{3}{8}$ of an inch.

Locality. Cor. Crag, Sutton and Gedgrave.

Red Crag, Sutton.

I possess only a few specimens of this shell from each formation, which I have considered as distinct. It somewhat resembles *N. elegans*, but differs in several characters. The transverse ridges are more obtuse than in *N. elegans*, and almost invisible between the ribs, which are prominent and distant. The mouth has externally a thickened varix in the adult state, and the lip is much thickened within, with four or five obtuse dentations. The siphonal canal is deeply indented and reflected towards the back. It differs from the figure of *B. pusillum*, Phil. (t. 27, f. 15), in having the volutions more convex, and the costæ more distant, with a more circular aperture, and fewer dentations, and it may be considered intermediate between that species and *N. elegans*. It is more elongated and elegant than *B. graniferum*, Dujardin, and has the costæ further apart.

7. NASSA MONENSIS. *Forbes.* Tab. III, fig. 5.

NASSA MONENSIS. *Forbes.* Proceedings of Geol. Soc. vol. iv.

— PROXIMA. *S. Wood.* Catalogue 1842.

N. Testá elongato-ovatá, turritá, longitudinaliter costatá; anfractibus convexis, striatis; suturis profundis; costis 12; aperturá ovatá; labio unidentato; labro intus denticulato.

Shell elongato-ovate, turreted, longitudinally costated; whorls convex, rugosely striated; suture deep; costæ 10—12; aperture ovate, with a tooth at the upper part of the left lip; outer lip dentated within.

Axis, ⅜ of an inch.

Locality. Red Crag, Sutton.

This species is at present rare; I have met with but three specimens. It appears to be quite distinct from any shell with which I am acquainted, and is considered by Mr. Forbes to be identical with the shell above referred to, which was first found in the Pleistocene beds of the Isle of Man. It is not known as a living species.

8. NASSA CONGLOBATA. *Broc.* Tab. III, fig. 9.

BUCCINUM CONGLOBATUM. *Brocchi.* p. 334, t. 4, fig. 15, 1814.
— PUPA. *Brocchi.* p. 334, t. 4, fig. 14.
NASSA CONGLOBATA. *S. Wood.* Catalogue 1842.

N. Testá ovatá, subglobatá, inflatá, tumidá; transversim sulcatá; anfractu ultimo globoso, spirá abbreviatá, aperturá subrotundá; labio rugoso unidentato; basi profundè emarginatá; labro intus denticulato.

Shell ovate, subglobose, inflated, transversely striated, or sulcated; spire short; mouth ovate, contracted at the upper part; left lip expanded, rugose, with a tooth at the upper angle of the aperture; base short; canal reflected; outer lip dentate within.

Axis, 1½ inch.

Locality. Red Crag, Walton Naze.

This is, I believe, at present, an unique specimen. Mr. Charlesworth was the fortunate finder, and through that gentleman's liberality it now enriches my collection of Crag. The specimen is a little worn, but there is no doubt of its identity; I have some Italian specimens in the same condition. The artist has given rather too great a prominence to the ridges upon the surface. The outer lip is thickened externally, with about nine or ten teeth upon the interior, and a few obsolete denticulations upon the left or columellar lip, which was largely extended, though broken, in my specimen.

9. NASSA PRISMATICA. *Broc.* Tab. III, fig. 6.

BUCCINUM PRISMATICUM. (?) *Broc.* p. 337, sed non fig.
— *Phil.* En. Moll. Sic. vol. i, p. 220, 1836.
— *Nyst.* Coq. foss. de Belge, p. 576, pl. 43, fig. 12, 1844.
BUCCINUM ELEGANS. *Dujard.* Trans. Geol. Soc. of France, 1837, p. 298, pl. 20, fig. 3, 10.
NASSA MICROSTOMA. *S. Wood.* Catalogue 1842.

N. Testá ovato-conoideá, longitudinaliter costatá, striis elevatis transversis; anfractibus rotundatis; suturis profundis; labio supernè uniplicato, basi reflexá emarginatá; labro extus incrassato, intus denticulato.

Shell ovato-conical, longitudinally costated; costæ 15 on the last volution, covered with elevated transverse striæ; whorls rounded and suture deep; one tooth on the upper part of the left lip; aperture subcircular; outer lip thickened, denticulated within.

Axis, ⅞ of an inch.

Locality. Cor. Crag, Gedgrave.

Red Crag, Sutton. Recent, Mediterranean.

When my Catalogue was compiled I possessed but two or three specimens from the Red Crag, and those not in very good condition. A few have lately been obtained from the Coralline Crag, at Gedgrave, by Mr. Daniels and myself. These are in better preservation, and sufficiently perfect to bear a fair comparison, and may be considered as identical with the recent species from the Mediterranean. They appear to be the same as the *Buc. elegans* of Dujardin, some specimens of which, from Touraine, were given to me by Mr. Lyell. It is different from the figure of *B. prismaticum*, Broc., Table v, fig. 7, as pointed out by Philippi.

10. NASSA RETICOSA. *J. Sow.* Tab. III, fig. 10, *a—h.*

NASSA RETICOSA,	var. *a.*	VULGARIS.	*S. Wood.* t. 3, fig.	10, *a.*
—	var. *b.*	RUGOSA	-	-	10, *b.*
—	var. *c.*	ELONGATA	-	-	10, *c.*
—	var. *d.*	THIARA	-	-	10, *d.*
—	var. *e.*	FENESTRALIS	-	-	10, *e.*
—	var. *f.*	DEFORMIS	-	-	10, *f.*
—	var. *g.*	CONCINNA	-	-	10, *g.*
—	var. *h.*	COSTATA	-	-	10, *h.*
—	var. *i.*	PERVERSA, t. 19, fig. 13, *a—b.*
BUCCINUM RETICOSUM.		*J. Sow.* Min. Conch. t. 110, fig. 2, 1815.
—	RUGOSUM	-	-	-	-	110, fig. 3.
—	ELONGATUM	-	-	-	-	110, fig. 1.
—	—	*Nyst.* Coq. foss. de Belg. p. 575, pl. 45, fig. 1, 1844.
—	—	var. *a.*	-	-	-	-	45, fig. 2.
—	—	var. *b.*	-	-	-	-	45, fig. 3.
		Dale. Hist. and Antiq. of Harwich, pl. 10, fig. 12, 1730.

N. Testá elongatá, costatá, reticulatá, rugosá, granulatá, striatá vel sulcatá; spirá elevatá, aliquando brevi; aperturá ovatá; basi reflexá, emarginatá; labro intus denticulato.

Shell very variable, sometimes costated, rugose, granulated, reticulated, striated, or sulcated; spire generally elevated; aperture ovate, with a short emarginate reflected base; outer lip denticulated within.

Locality. Red Crag, *passim.*

This is one of the most abundant shells in the Red Crag, and specimens of one or more of the varieties may be procured wherever a section of that formation is visible. *B. reticosum*, Sow., is here considered as the type of the species, as they are all more or less reticulated, and not all elongated. In some of the varieties the outer lip is thickened and denticulated within, while, in others, it is thin and plain. I have not been able to identify *N. reticulata* with any of these varieties, and have not as yet seen an undoubted specimen of that species from any of the three divisions of the Crag formation.

The duration of its existence appears to have been limited to the Red Crag period, as I am not aware of its having been found elsewhere, excepting in the contemporaneous formation in Belgium. The reversed variety, Table XIX, fig. 13, is from the cabinet of Mr. Gibson. It was found at Walton-on-the-Naze, by the late Rev. Mr. Rogers, of Lackford, in 1820, and is the only specimen I have ever seen. Another very short variety is in my cabinet, but much worn.

BUCCINUM,* *Linn.* 1767.

TRITONIUM. *Müller,* 1774.

Gen. Char. Shell subovate or ovato-conical, generally ventricose; spire slightly elevated; apex rather obtuse; aperture large and oval, with an emarginate base; outer lip simple, sometimes sinuated, and without denticulations; inner lip slightly expanded, columella smooth. Operculum corneous.

The common *Buc. undatum* may be considered as the type of this genus, which intervenes between and connects Nassa on the one side, with Trophon on the other. The close resemblance of the animal of *Trophon* (*Fusus*) *antiquum* to that of the common whelk has induced some conchologists to unite them in one genus. The short and emarginate base is here considered sufficient to separate these shells from *Trophon*, and the large open naked aperture from *Nassa*.

1. BUCCINUM DALEI. *J. Sow.* Tab. III, fig. 10, *a—d.*

     BUCCINUM DALEI.   *J. Sow.* Min. Conch. t. 486, fig. 1, 2, 1825.

         —   OVUM.   *Turt.* Zool. Journ. vol. ii, p. 366, pl. 13, fig. 9, 1827.

         —   DALEI. *Nyst.* Coq. foss. de Belg. p. 570, 1844.

         —   CRASSUM.   -   -   -   p. 569, t. 44, fig. 7.

         —   DALEI. *S. Wood.* Catalogue 1842.

   HALIA FLEMINGIANA.   *Macgill.* Moll. Aberd. p. 189 (fide Jeffreys).

                  Dale. Hist. and Antiq. of Harwich, pl. 10, fig. 3, 9, 1730.

B. Testá ovato-conicá, ventricosá, striatá, crassá, apice obtuso; anfractibus convexiusculis, infernè dilatatis, suturis profundis; aperturá ovatá; canali brevi; labro acuto, intùs incrassato.

Shell ovato-conical, somewhat ventricose, transversely striated, thick and strong, with an obtuse apex; volutions somewhat flattened on the upper half, dilated on the lower part; aperture large and ovate, with a wide and short canal; outer lip sharp and plain, thickened inwardly, and a broad expanded inner lip.

Axis, from 1 to $2\frac{1}{3}$ inches.

Locality. Cor. Crag, Ramsholt and Gedgrave.

         Red Crag, Walton Naze and Sutton.   Recent, British Sea.

This species is exceedingly abundant at Walton-on-the-Naze, where the specimens are generally in a high state of preservation, but it is rarely found in the Coralline Crag. It is very variable in the proportionate dimensions of the aperture, some having it half the length of the axis of the shell, while in others it is two thirds, and in the young shell

---

* Etym. *Buccinum,* a trumpet, or fish with a shell like a trumpet.

it is particularly large. In some specimens the striæ are very conspicuous, while in others they are somewhat indistinct; they may, however, be traced in every specimen. The left lip extends some way over the body of the shell, and there is a prominent ridge on the edge of the columella, as pointed out by Mr. J. Sowerby in ' Min. Conch.,' corresponding in these characters with the genus Nassa, but it has a large and open aperture, with a plain and simple outer lip, which is free from denticulations. It may be considered as a rather aberrant form connecting the two genera.

Buc. ovum, Turt., has hitherto been represented and described as a perfectly smooth shell, and considered, in consequence, as a new species, distinct from *B. Dalei*, with which it agrees in all its other characters. Professor E. Forbes, however, has informed me that Mr. M'Andrew has detected some faint traces of striæ upon a specimen of *B. ovum* in his possession. It may then, I think, be fairly presumed as identical with *B. Dalei* of the Red Crag, and its extreme rarity be looked upon as the dying out of a species once exceedingly abundant in this country; and in thus becoming extinct, may not its want of, or rather its faint and imperfect striæ be from failure in vigour in those organs necessary for such distinction, consequent upon the approaching demise of the species?

2. BUCCINUM UNDATUM. *Linn.* Tab. III, fig. 12, *a—d.*

<blockquote>
BUCCINUM UNDATUM.   *Linn.* Syst. Nat. p. 1204, 1766.

—   STRIATUM (?).   *Penn.* Brit. Zool. iv, p. 121, 1776.

—   UNDATUM.   *Brown.* Illust. Brit. Conch. pl. 49, fig. 8, 1827.

—   ANGLICANUM (?).   -   -   -   pl. 49, fig. 11.

—   TENERUM.   *J. Sow.* Min. Conch. t. 486, fig. 3, 1825.

—   —   *Nyst.* Coq. foss. de Belg. p. 571, pl. 43, fig. 9, 1844.

—   UNDATUM.   *S. Wood.* Catalogue 1842.

TRITONIUM UNDATUM.   *Lovén.* Ind. Moll. Scand. p. 12, 1846.

    Dale. Hist. and Antiq. of Harwich, pl. 10, fig. 8, 1730.
</blockquote>

B. Testá ovato-conicá, ventricosá, striatá, costatá, undulatá; anfractibus convexis; aperturá ovatá, labro sinuato, canali brevissimo aperto.

Shell ovato-conical, ventricose, generally thin, with, sometimes without, undulate costæ; whorls convex; aperture ovate, with a thickened and slightly sinuated outer lip; base emarginate; canal short and open.

Axis, 3 inches.

Locality.   Cor. Crag, Ramsholt.

           Red Crag, Butley, Newbourn, Sutton, and Walton.

           Mam. Crag, Bridlington.   Recent, British and North Seas.

This shell first appears in the Coralline Crag, where, however, it is by no means abundant. In the Red Crag it becomes more so; but the greater number of the specimens, being much reduced in substance, are fragile, and difficult to obtain; the outer coating having been in some instances entirely removed, its peculiarly striated markings are thereby obliterated. Some specimens so precisely resemble the common recent shell that I have no doubt of its identity. Var. *a*, probably *B. striatum* Pennant, appears to be rather of an aberrant form in this species; the outer lip is similarly sinuated, but not periodically thickened, and the consequent absence of the undulatory elevations upon the body of the shell gives a great apparent difference to this variety. The figures above referred to are of course the extreme forms, but I think they can be fairly connected by the series I possess. The proportionate dimensions of the aperture, with regard to the axis of the shell, is a character by no means constant.

PURPURA,* *Adanson*, 1757.
BUCCINUM (spec.) *Linn.*

Gen. Char. Shell ovate, thick, and strong; spire generally short, sometimes elongated; external surface striated, sulcated, muricated, granulated, or tuberculated; aperture ovate, somewhat dilated; outer lip crenated, occasionally dentated within; columella usually flattened; base subcaniculate or emarginate. Operculum corneous, with a lateral nucleus.

Shells of this genus are generally very thick and strong, and may be distinguished from Nassa or Buccinum by a peculiar flatness upon the columella, while they resemble those genera in many other characters. The canal is short, straight, and not reflected; but the animal is said to present several constant characters, by which it may be considered as distinct.

It is not yet known as an Eocene fossil, though M. Deshayes speaks of it as from some beds of the Oolitic period.

1. PURPURA LAPILLUS. *Linn.* Tab. IV, fig. 6, *a—h.*

PURPURA LAPILLUS, var.	*a*, CRISPATA.	*S. Wood.* t. 4,	fig. 6, *a.*	
—	*b*, BREVIS	-	-	fig. 6, *b.*
—	*c*, ELONGATA	-	-	fig. 6, *c.*
—	*d*, VULGARIS	-	-	fig. 6, *d.*
—	*e*, ANGULATA	-	-	fig. 6, *e.*†
—	*f*, CARINATA	-	-	fig. 6, *f.*
—	*g.* IMBRICATA	-	-	fig. 6, *g.*
—	*h*, INCRASSATA	-	-	fig. 6, *h.*
—	*i*, COMPRESSA, t. 19, fig. 12, *a, b.*			

BUCCINUM LAPILLUS. *Linn.* Syst. Nat. p. 1202, 1766.
TRITONIUM LAPILLUS. *Müll.* Zool. Dan. Prod. 244, 1776.
PURPURA LAPILLUS. *Lam.* An. sans Vert. tom. vi, 1815.
BUCCINUM CRISPATUM. *J. Sow.* Min. Conch. t. 413, 1823.
— INCRASSATUM. - - - t. 414, fig. 2.

* *Purpura*, "the shell-fish from which purple is taken." The celebrated dye of the ancients is now considered to have been produced from the *Murex trunculus*, a shell common in the Mediterranean, and not from the *Purpura lapillus*, which is not known in that part of the world.

† This is from the cabinet of Mr. Lyell.

Murex angulatus. *Woodward.* Geol. of Norf. t. 3, fig. 23, 1833.
— elongatus. - - - - - - fig. 22.
— bulbiformis - - - - - fig. 21.
— lapilliformis - - - - - fig. 25.
— compressus - - - - - fig. 26.
Purpura lapillus. *Flem.* Brit. An. p. 341, 1828.
— *Gould.* Invert. of Massachusetts, p. 301, 1841.
— *Thorpe.* Brit. Mar. Conch. p. 212, 1844.
— *S. Wood.* Catalogue 1842.
— *Morris.* Catal. of Brit. Foss. p. 160, 1843.
Murex incrassatus. *Nyst.* Coq. foss. de Belg. p. 548, pl. 43, fig. 2, 1844.
Dale. Hist. and Antiq. of Harwich, tab. 10, fig. 4, 1730.

P. Testá ovatá, vel elongatá, ventricosá, crassá, transversim sulcatá, longitudinaliter imbricatá; apice acuminato; anfractibus 6—8 convexis; aperturá ovatá; labro acuto; columella planiuscula; canali brevi.

Shell variable, ovate, elongate, ventricose, angulose, compressed, thick, and strong, transversely sulcated or ridged, longitudinally imbricated; whorls 6—8, convex; outer lip sharp, sloping inwards where it is thick and denticulated, with a broad and flattened columella; canal short, and a small but open umbilicus.

Axis, reaching $2\frac{3}{4}$ inches.

Locality. Red Crag, *passim.*

Mam. Crag, Bramerton and Thorpe.—Recent, Britain and North America.

This is one of the most abundant fossils of the Red Crag, and may be obtained wherever a section of that formation is visible. It is exceedingly variable, and no dependence can be placed either upon proportionate dimensions or upon the angle of volution. In some specimens the aperture measures two thirds the entire length of the shell, while in others it is not more than two fifths. I have little doubt but the shells figured by Woodward, above referred to, are deformed specimens of one species produced by the same cause, to which I have assigned, in my Catalogue, the many different varieties of the *Littorina littorea*. Dr. Gould appears to have divided the American shells into two groups, those which are smooth and thick, as the true *lapillus*, and the imbricated ones, which have a rather longer canal, and were considered by Lamarck as a distinct species, under the name *P. imbricata*. The surface of all the specimens from the Crag, when in good condition, are covered with imbrications, which induced the author of ' Min. Conch.' to consider it distinct; but the recent British species is found sometimes with quite as rough a surface, and specimens in my cabinet are so precisely similar to some of the fossils that, except in colour, they could not be distinguished. The thick and ponderous variety (fig. 6 *h*), with rather more elevated and distinct carinæ, figured and described in 'Min. Conch.' as *P. incrassata*, is rather an aberrant form in this species, but may be connected by a large series of specimens; and several able conchologists, who have examined the varieties I possess, are of the same opinion, that it is only an extreme form. In the Crag seas it appears to have attained a greater magnitude than the generality of recent specimens, either from the British seas or from those of America, and also to have exhibited a greater range in variation; although Dr. Gould remarks, in his very able descriptions of the Invertebrata of Massachusetts, that scarcely two specimens can be found alike on that shore, where it is abundant. It is quite a littoral species, and is found upon rocks up to high-water mark. In the Mammaliferous Crag it is associated with estuary species. It first appears in England in the Red Crag, which seems to have been peculiarly favorable to its development; and, as Professor E. Forbes suggests, in his 'Report upon the Geological Relations of the Existing Fauna and Flora of the British Isles,' p. 93, is probably of American origin. Mr. Lyell speaks of it as a fossil of that country, in his paper upon the Miocene Tertiary Strata of Maryland, &c.

2. PURPURA TETRAGONA. *J. Sow.* Tab. IV, fig. 7, *a—d.*

PURPURA TETRAGONA. *J. Sow.* Min. Conch. t. 414, fig. 1, 1823.
MUREX ALVEOLATUS.   -   -   -   - t. 411, fig. 2, 1823.
—   *Nyst.* Coq. foss. de Belg. p. 547, pl. 43, fig. 1, 1844.
PURPURA TETRAGONA. *S. Wood.* Catalogue 1842.

P. Testá ovato-ventricosá, vel elongatá, subfusiformi; transversim sulcatá, sulcis latè profundis, decussatis; anfractibus supernè planatis; aperturá ovatá; canali rectiusculá; columellá subperforatá; labro intus denticulato.

Shell ovate, ventricose, sometimes subfusiform, and elongate; sulcated and decussated, forming large and deep alveoli upon the exterior; volutions rather flattened above, subcarinated; aperture ovate, with a short and open canal, a little inclined backwards; columella subperforated; outer lip dentated within.

Axis, 1½ inch.

Locality. Red Crag, Sutton and Walton Naze.

This shell appears to have had a very limited vertical range as far as it is at present known. I have met with it only in the Red Crag, where it is exceedingly abundant, and very variable in its proportionate dimensions. *Murex alveolatus,* Sowerby, is, I imagine, only an elongated variety of this species, as the two may be connected by every intermediate form between the extremes figured (fig. 7, *a—b*), by the suite of specimens in my own cabinet. The only difference is, a greater prolongation of the convoluted cone, as the markings are precisely similar, and the aperture of the same form. It has four or five very elevated and rounded transverse ridges, with one, or sometimes three, intermediate striæ; these are decussated by the raised and subfimbriated edge of the outer lip, thereby producing upon the exterior the deep alveoli which so greatly distinguish this shell. The left lip is thin, with the columella a little flattened; and in the adult shell the outer lip has about half a dozen denticulations on the inside; the flexure of the canal leaves the umbilicus partly open. Fig. 7 *d* is, I presume, a variety of this species, which has lost a portion of its outer coating, and is thinner, much after the manner in which the substance of the shell is reduced in the var. *tenerum* of *Buccinum undatum.*

MUREX, *Linn.* 1767.
CHICOREUS. *Montf.* 1810.
TRITON. *Flem.* 1828.
PHYLLONOTUS. *Swains.* 1840.
PTERONOTUS. - - -
MURICIDEA. - - -

Gen. Char. Shell turreted, elongato-ovate, or fusiform; generally ventricose, thick, and strong; externally rugose, with three or more longitudinally branched, spinous, fringed, or reflected fimbriæ, more or less elevated, foliaceous, or tuberculous; spire prominent and acute; aperture ovate, terminating at the base in a partly-closed canal; operculum corneous, concentrically formed and pointed.

The shells of this genus have three or more divisions of the volution, formed by an extended or fimbriated margin of the outer lip, at different periods of growth, by which means the shell is elegantly ornamented externally; and among the recent portion of the genus are some of Nature's most beautiful productions. It differs from Ranella, which has only a bipartite division of the whorl, and in Triton the varices are more irregular, and not fimbriated. This genus is found in the Oolite, and twenty species are enumerated as belonging to the Paris basin. M. Michelotti mentions forty-four from the Subapennine beds. Only two species have as yet been found in the Crag, and those are by no means numerous in individuals.

1. MUREX ERINACEUS, *Linn.*

MUREX ERINACEUS. *Linn.* Syst. Nat. p. 1216.
— *Pennant.* Brit. Zool. iv, t. 76, fig. 95, 1776.
— *Brocchi.* Conch. foss. Subap. p. 391, pl. 7, fig. 11, 1814.
— *Dujardin.* Mém. de la Soc. Géol. de France, 1837, p. 295.
— *S. Wood.* Catalogue 1842.
TRITON ERINACEUS. *Flem.* Brit. An. p. 356, 1828.

M. Testá ovato-fusiformi, transversim sulcato-rugosá, quadrifariàm ad septifariàm varicosá; varicibus elevatis; aperturá ovatá; canali clauso recurvá.

Shell subfusiform, ovate; roughly sulcated transversely with from four to seven rather obtuse varices; aperture ovate, with a slightly recurved canal, which is outwardly closed in the adult state.

Locality. Mam. Crag, Bramerton. Recent, Britain.

One specimen of this species was found some years since by Mr. J. Wigham, of Norwich, and sent to Mr. Lyell, with whom I examined it, and we were both satisfied of its identity with the well-known recent shell. Upon application to the owner of the specimen for the privilege of having it figured, he informed me it had been sent to Mr. Green, of Bacton, for the purpose of publication, and unfortunately lost.

2. MUREX TORTUOSUS. *J. Sow.* Tab. IV, fig. 9.

MUREX TORTUOSUS.   *J. Sow.* Min. Conch. t. 434, fig. 2, 1823.
      —   *S. Wood.* Catalogue 1842.
      —   *Nyst.* Coq. foss. de Belg. p. 545, pl. 41, fig. 14, 1844.

M. Testá trigoná, fusiformi; transversim latè sulcatá; tribus varicibus angulosis, foliacis, ornatá; interstitiis nodulosis; aperturá subquadratá; canali non clauso, elongato, rectiusculo.

Shell turreted, fusiform, trigonal, with three sharp varices in each volution, intermediate spaces nodulous, and with elevated transverse ridges; aperture ovate; canal contracted, but open.

Axis, 1¾ inch.

Locality.   Cor. Crag, Sutton.
              Red Crag, Sutton and Walton-on-the-Naze.

A rare shell in the Red Crag, but more so in the Coralline beds, in which I have met with only a few fragments. It differs from *M. erinaceus* in having three elevated, sharp, reflected, rather tortuous, and slightly fimbriated varices, in one volution. Upon the spire, or in its young state, these are not so distinct; they are then numerous, sharp, and elevated: as it grows older, the intermediate ones dwindle into obtuse knobs, which are seen between the varices. It is rugosely striated transversely or along the volutions, with four or five elevated ridges, the upper one of which is the most prominent, giving an angular form to the volution; it has fine striæ between the ridges, which are seen only in those specimens that are well preserved. Traces of obtuse denticulations within the outer lip may be seen in some specimens.

TRITON,* *Montf.* 1810.
TRITONIUM.   *Cuv. Phil.*

Gen. Char. Shell ovate or fusiform, thick and strong, carinated, striated, or tuberculated; with discontinuous varices placed at irregular distances; spire more or less elevated, with a rather obtuse apex; aperture subcircular, or ovate, terminating in a generally elongated, and slightly curved, or raised canal; columella lip granulated or plicated; outer lip thickened and reflected, often dentated within; and in a living state it has a corneous operculum.

This genus appears to be intermediate between Murex and Ranella, the former of which has three or more imbricated varices in each volution, while in the latter there are only two, which are regular and obtuse. In this genus they are obtuse and irregular; sometimes with only a thickened margin to the aperture, formed when the shell has attained its full dimensions. Most of the recent species of this genus are natives of tropical regions, though several are found living in the Mediterranean. It first appears in the Eocene period, and has greatly increased in number of species in the recent state.

* Etym. *Triton,* one of the sea deities.

1. **Triton heptagonum.** *Broc.* Tab. IV, fig, 8.

Murex heptagonus. *Brocchi.* Conch. foss. Subapenn. vol. ii, p. 404, t. 9, fig. 2, *a—b*, 1815.

T. Testá turritá, ovato-conicá subfusiformi; apice obtuso, anfractibus angulatis, prismaticis, supernè tumidis prope suturam depressis, plano-canaliculatis; sulcis transversis crenatis; varice solitaria marginali; aperturá ovatá; labro intus dentato; canali aperto vix recurvo.

Shell turreted, subfusiform, with an obtuse apex; volutions obtusely angulated; the upper part tumid and convex, with a flattened depression round the whorl near the suture; transversely ridged or sulcated, with distant longitudinal elevations; aperture ovate, and a thickened varix; outer lip denticulated within; a narrow but open canal of moderate length, slightly recurved.

Axis, $1\frac{3}{4}$; *transverse diameter,* $1\frac{1}{8}$ inch.

Locality. Cor. Crag, Gedgrave.

A single specimen of this beautiful shell is all I have as yet seen. It graces the cabinet of Stephen Perry, Esq., of Rushmere, who has kindly intrusted me with it for illustration.

It differs slightly from the figure by Brocchi in having a shorter canal, and a less number of ridges, and the outer lip is less angular than in the Italian specimen, but these differences are probably only local, and I have no doubt of its identity. The volutions are slightly prismatic, with about seven faces. There are two varices in this specimen, but it is probably an overgrown individual, and the last half volution an effort of growth beyond its usual size; the edges project beyond the margin of the aperture, giving the varix a slightly fimbriated character, and the figure thus represents it, but these hollow fimbriæ are produced by erosion or decomposition of the shell at that part. It has a small tooth-like projection upon the body whorl at the upper part of the aperture, forming there a shallow canal, and a few folds upon the lower part of the columella, which are probably only the elevated ridges of the exterior imperfectly covered by the left lip.

PYRULA,* *Lam.* 1801.

Sycotypus. *Brown.* 1756 (fide J. E. Gray).

Ficula. *Swainson.* 1840.

Gen. Char. Shell thin, subovate, ventricose, fig, or pear-shaped, generally striated or cancellated upon the exterior; spire short and depressed, consisting of few volutions; aperture large and wide, terminating in a long, narrow, open canal; columella smooth, slightly tortuous; outer lip sharp; inner lip very thin and expanded.

This genus still contains a large number of shells not possessing the above characters, being thick and heavy, with a polished and sometimes a nodose exterior. Swainson has justly separated these, and proposed the name of Ficula for the section we are now considering, and Lamarck intended those species that are known in commerce by

* Etym. The diminutive of *Pyrum,* a pear.

6

the name of Figs as the type of his genus Pyrula.   Mr. G. B. Sowerby, in his Genera of Recent and Fossil Shells, had also previously excluded all others from this genus, for which he retained the name of Pyrula; and although the dissimilar shells still united with these must hereafter be separated, the name of Pyrula ought to be retained for this section.   The recent species of this genus, or section, are found only within the tropics.

1. PYRULA RETICULATA. *Lam.* Tab. II, fig. 12.

BULLA FICUS, *var.* 1.	*Broc.* Conch. foss. Subap. p. 279, 1814.
PYRULA RETICULATA.	*Dujard.* Mém. de la Soc. Géol. de France, Ency. Méth. pl. 432, fig. 2.
—	*Lam.* An. sans. Vert. 2d edit. t. ix, p. 510.
—	*S. Wood.* Catalogue, An. and Mag. Nat. Hist. 1842, p. 543, pl. 5, fig. 17.
—	*G. Sowerby.* Genera of Shells, fig. 1.
—	*Phil.* En. Moll. Sic. vol. ii, p. 180, 1844.

P. Testá ficoideá, pyriformi, ventricosá, tenui, cancellatá; spirá brevissimá, convexá, retusá; anfractibus circa quatuor inflatis; aperturá amplá, ovatá; canali angustatá; labro acuto.

Shell thin, ventricose, pear or fig-shaped; spire short and convex; volutions about four, cancellated, transverse striæ the more elevated; aperture large, subovate; with an elongated caudal termination; outer lip sharp.

Axis, 3 inches.

Locality.   Cor. Crag, Ramsholt.   Recent, Indian Ocean.

My cabinet contains only two specimens of this shell, which appear to differ in some slight degree from the Oriental species.   The upper part of the outer lip is more elevated, the whole shell is not quite so slender, and the transverse ridges are rather broader and flatter.   The specimen figured does not represent the canal so elongated as it is in the recent specimens, but a portion of it is broken away; the lines of growth indicate a similar length.

If this be *Pyrula reticulata* of Lamarck, which I presume it is, its retirement or migration to the southward and eastward might have been through the seas that deposited the Touraine beds, whence, in all probability, it originally came, and had an extension of existence through the more modern (?) deposits of Calabria, in which it is found fossil, as quoted by Philippi.   Conceiving a communication to have then existed between the Mediterranean and the Arabian Gulf, a further extension to its present habitat might have taken place through the channel now so effectually closed by the Isthmus of Suez.

The presence of this species in a latitude so high as that of England has been accounted for upon the supposition that the temperature of the sea by which the Coralline Crag was deposited was more favorable to its existence than the seas of the same latitude are at the present day; but a very elevated temperature does not appear to have been essential to its existence, if we may judge from its associates, *Tricotropis borealis, Nucula pygmæa*, &c., which are now found only on our own northern coasts, and in the Arctic Seas.   The supposition that those northern forms at that time inhabited a deep-water portion of the Crag sea, is equivalent to a diminished temperature; while this tropical representative must have frequented a shallower portion, having a more elevated temperature, that is, if the law be admitted that lines of depth are equivalent to zones of latitude, or isothermal lines. This theory will not satisfactorily explain how these arctic and torrid representatives are quietly reposing together in the Crag beds without the intervention of disturbing causes, of which there are certainly no indications in the locality from which they were obtained. Whatever may have been the temperature of the Coralline Crag sea,—and I think it may have been rather more elevated than that of our present seas,—it is evident that these animals have now retired or migrated into those parts of the world, the one north and the other south, where the temperature of both is very different from that which must have been favorable to their existence at the period spoken of, and they have, therefore, in some degree changed their nature in assimilating such extremes to their present existence. Their mode of dispersion was, it is presumed, by means of currents, which perhaps had at that period a northerly direction, thus dispersing those species which are now considered as arctic forms, while the torrid representatives might have died out where they are now found, and their dispersion to the southward may have been by southerly currents from the contemporaneous seas, producing the Touraine beds. Conclusions regarding the temperature of the sea during the period the Crag was deposited have been drawn from the presence of such animals as Pyrula, Pholadomya, &c., and an elevation considerably above that of the British seas of the present day has been assigned to it at that period in consequence, while I believe the change only to have taken place in the animals themselves; and this might arise from their acquiring habits of enduring increased or diminished temperature by gradual migration, until extended ·location had caused them to reach such extremes as would have been fatal to their existence had their removal been suddenly effected.

TROPHON,* *De Montf.* 1810.
FUSUS (spec.) *Bruguière.*
TRITONIUM. *Müller.* 1773.
ATRACTUS. *Agassiz.* 1840.

Gen. Char. Shell fusiform and turreted, sometimes ventricose, with many rounded volutions, costated, rarely smooth, often striated; aperture terminating in a moderately elongated canal; outer lip simple; columella smooth; operculum corneous.

As the name of *Fusus* may be more correctly reserved for those species of which *Murex porrectus*, Brand., would form the type, I have used the above, proposed by De Montfort, 1810, and adopted by Möller, 1842, in preference to *Tritonium*, Müller, 1773; the latter being objectionable, on account of two names of similar import being extensively known as generic terms in the same class, namely, Tritonia and Triton.

* Etym. ?

Shells with an elongated canal, which properly constitute the genus *Fusus,* are characteristic of a tropical climate.   The greater part of the species of this genus from the Crag are northern forms.

1. TROPHON ANTIQUUM. *Müll.* Tab. V, fig. 1, *a—k.*

> TRITONIUM ANTIQUUM.   *Müll.* Zool. Dan. tab. 118, fig. 1-3, 1773.
> MUREX CONTRARIUS.   *Gmel.* Syst. p. 3564, 1788.
> FUSUS ANTIQUUS.   Ency. Meth. pl. 426, fig. 5.
> — DESPECTUS (?).   -   pl. 426, fig. 4.
> — CONTRARIUS   -   pl. 437, fig. 1, *a* 6.
> MUREX DESPECTUS.   *Mont.* Test. Brit. p. 256, 1803.
> — ANTIQUUS (?).   -   -   - p. 257.
> — STRIATUS.   *J. Sow.* Min. Conch. t. 119, 1815.
> — *var.* CARINATUS.   -   -   - t. 22, 1813.
> — CONTRARIUS.   -   -   - t. 23.
> BUCCINUM CONTRARIUM.   *G. Sow.* Genera, fig. 4.
> FUSUS CONTRARIUS.   *Phil.* En. Moll. Sic. vol. ii, p. 179, 1844.
> —   —   *Nyst.* Coq. foss. de Belg. p. 500, pl. 41, fig. 1.
> — ANTIQUUS.   *Lam.* An. sans Vert. (2d edit.) ix, p. 477, 1845.
> — CARINATUS   -   -   -   -   - p. 449.
> — CONTRARIUS   -   -   -   -   - p. 462.
> — SINISTRORSUS (?).   -   -   -   - p. 374.
> TRITONIUM ANTIQUUM.   *Lovén.* Ind. Moll. Scand. p. 11, 1846.
> *Dale.* Hist. of Harwich, pl. 10, fig. 5-6, 1730.

Tr. Testá crassá, turritá, fusiformi, sulcatá, carinatá, vel tenuissime striatá; apice papilliformi, anfractibus valdè convexis, tumidis; aperturá ovatá; labro simplici, intus lævigato; canali brevi, emarginatá.

Shell variable, strong, thick, ponderous, fusiform, turreted, sulcated, bicarinated, or finely striated, with a white, naked, and mammillated apex; whorls convex, tumid; aperture ovate; lip simple, smooth within; canal short.

Axis, 6 inches.

Locality.   Red Crag, *passim.*

Mam. Crag, Bramerton and Bridlington.   Recent, Britain.

I have very little doubt of the shell found so abundantly in the Red Crag, with its sinistral volutions, being a variety of the common *Tr. antiquum.* This species is exceedingly variable, and no dependence can be placed upon any proportional dimensions of its aperture, as may be seen from the extremes of variation figured in Tab. v, and between which every intermediate form may be procured.   Some specimens have an aperture exceeding two thirds of the axis, while in others it is not more than one third.   The smooth and obtuse state of the young shell, at the apex of perfect specimens, is a good distinguishing character in this species; but in the greater number of the fossil specimens that part is destroyed, excepting in the striated variety, so common at Walton-on-the-Naze.   The Belgian shell, figured by M. Nyst, is an extreme variety of this species.   I have never seen the canal so much produced in the English specimens;

and that author speaks of his shell as pointed at the apex : " Sa spire allongée et pointue au sommet se compose de sept tours très convexes, ordinairement lisses dans les individus fossiles de l'Angleterre." None of my specimens have that character, and, in fact, I have not any specimen that is quite smooth, although many are much rubbed and eroded; traces of striæ may be detected in all that are in my cabinet. Although specimens of the dextral variety, precisely similar in sculpture to that of the recent shell, are not uncommon in the Red Crag, and it may be worthy of remark, that the general character of this variety is a bicarinated or tricarinated form, while the sinistral one is very rarely so, but resembles the finer striæ of the recent shell. This, as it is well known, is one of the most abundant shells in the Red Crag, and at least one of its varieties may be found wherever a section of that formation is visible; a good series may therefore be supposed in the hands of every collector. Prof. E. Forbes, in his 'Report upon the Geological Relations of the Existing Fauna and Flora of the British Isles,' has separated these varieties, and considers the carinated one as a distinct species. I am not well acquainted with the recent shell in all its variations, but I believe the Crag varieties, above described, to belong to one species. My cabinet contains every possible form of striation, graduating from the carinated ones unto those possessing the finest striæ; but I have not yet seen a Crag specimen with undulations. The extension of variation here given to this species is not more, or even so much, as is allowed by several eminent conchologists who admit the probability of the *Purpura incrassata* being a variety of the *P. lapillus.*

M. Philippi (En. Moll. Sic. p. 179) enumerates this species among the Sicilian fossils, which is in all probability identical with the shell now found living in the Mediterranean. This is considered a distinct species by M. Deshayes, and named by him *Fusus sinistrorsus.* On a comparison I have made with some recent specimens in the British Museum, and also with a fossil specimen from Palermo, in the Museum of the Geological Society, I could not detect any character by which it might be considered specifically distinct. The sculpture is different from the recent British specimens; but, as far as the shell alone can determine the species, it does not differ from some of the Crag specimens, and they can be connected with the other varieties.

Prof. E. Forbes, in his Report above alluded to, speaks of the dextral variety of this species as its normal condition; I am rather inclined to think otherwise, and that the sinistral variety was the original form of volution, and not a monstrosity; and that it has died out in the northern seas, and been replaced by the dextral form, while the sinistral one has retired to the southward. The left-handed specimens, now occasionally found in the British seas, may possibly be a remnant of this race; but I am more inclined to believe that they are merely monstrosities of the dextral variety, such as are sometimes found among other species, and are probably produced by an inversion of some of the important viscera

The carinated variety (f. 1 *k*) of *T. contrarius* is figured from a specimen belonging to Mr. Bean; it was found in the Mammaliferous Crag of Bridlington.

2. TROPHON ELEGANS. *Charlesworth.* Tab. VI, fig. 2.

    ATRACTODON ELEGANS. *Charlesworth.* Mag. Nat. Hist. New Series, 1837, p. 219, fig. 23.
    FUSUS (?) ELEGANS. *S. Wood.* Catalogue 1842.

Tr. Testá crassá, ovato-fusiformi; apice obtuso, transversim striatá, seu sulcatá; anfractibus sex convexis; suturis distinctis; aperturá ovatá; labro simplici, intus lævigato; labio supernè crasso dente munito; canali brevi, recurvá.

Shell thick and strong, ovato-fusiform, regularly and rather coarsely striated transversely; whorls six, convex, with a distinct suture; aperture ovate; outer lip plain and simple; inner lip with a tooth or callosity at the upper part, upon the body whorl.

Axis, 3 inches.

Locality. ———— ?

The shell from which the above figure is taken graces the cabinet of Mr. Robert Fitch, of Norwich, and is, I believe, an unique specimen. In its general outline, and in the form of the aperture, it appears to belong to this genus, and probably is not far removed from *Tr. antiquum.* It is covered with regular striæ, which become coarser towards the base of the shell. The specimen has been a good deal rubbed, especially about the apex, which is now very obtuse; and a part of the outer lip is broken away, as well as a portion of the canal, so that the lines of growth indicate a greater length than is now represented; the outer lip is very slightly sinuated, like that of *Tr. antiquum.* The callosity at the upper part of the left lip is similar to that observable in many shells of the Buccinoid family, where a thickening of the left lip, near the upper angle of the aperture, forms a sinus for the exit of the water, after having aerated the branchiæ, and the canal there is more effectually formed for that purpose.

This shell was found on the beach at Felixton. It is highly tinged with the oxide of iron, and is probably from the Red Crag.

3. TROPHON GRACILE. *Da Costa.* Tab. VI, fig. 10, *a—c.*

    BUCCINUM GRACILE. *Da Costa.* Brit. Conch. t. 6, fig. 5, 1778.
    MUREX ISLANDICUS. *Gmel.* Syst. p. 3555, 1788.
        —   *Chemn.* vol. iv, t. 141, fig. 1312, 13.
    MUREX CORNEUS. *Don.* Brit. Shells, pl. 38.
        —   *Mont.* Test. Brit. p. 158, 1803.
    FUSUS CORNEUS. *J. Sow.* Min. Conch. t. 35.
        —   *Nyst.* Coq. foss. de Belg. pl. 39, fig. 23, 1844.
    FUSUS ISLANDICUS. *Lam.* An. sans Vert. 2d edit. t. ix, p. 450.
        —   *Gould.* Inv. of Massach. p. 284, 1840.
    FUSUS ANGUSTIUS. *S. Wood.* Catalogue 1842.
    TRITONIUM GRACILE. *Lovén.* Ind. Moll. Scand. p. 11, 1846.
    BUCCINUM ANGUSTIUS. *List.* Conch. t. 913, fig. 5, 1685.
              Dale. Hist. of Harwich, pl. 10, fig. 7, 1730.

Tr. Testá turritá, elongato-fusiformi; anfractibus convexis, transversim striatis; aperturá ovatá; labro tenui, intus lævigato; canali elongatá recurvá.

Shell elongato-fusiform; turreted, with an obtuse apex; whorls 7—8, convex, striated along the volutions; outer lip sharp, slightly sinuous; aperture ovate, terminating in an elongated and curved canal.

Axis, 2⅛ inches.

Locality.　Cor. Crag, Gedgrave.

　　　　　Red Crag, *passim*.

　　　　　Mam. Crag, Bridlington.　　　　　　　　　Recent, British Seas.

Abundant in the Red Crag; and I have recently obtained two specimens from the Coralline beds.　The latter present a slight difference in sculpture from the recent specimens, having the striæ, or ridges, which cover the exterior of the shell, flatter and broader, with narrow sulci between them.　In the recent shell the striæ are narrow, sharp, and elevated, with broad spaces intervening.　Among a large series from the Red Crag these markings may be observed to vary; they are generally flat and broad, but in some they are sharp and narrow.　In many specimens the shell is left nearly smooth, with a deep depression at the suture, which is the principal difference between the fossil and the recent shell, and is probably produced by alteration or loss of a portion of the shell, as, in some instances, the volutions appear nearly separated.　The apex is obtuse and smooth, but can scarcely be said to be mammillated, and the canal is generally a good deal twisted.　Some specimens are short and tumid, with a diameter three sevenths of the axis, while others are elongated, with a diameter only one third.　The same proportionate variations may be observed in the recent shell.

A specimen, sent from Mr. Bean, with a label " *F. Listeri*, Crag, Bridlington," appears to me only a variety of this species.

4. TROPHON ALTUM. *S. Wood*. Tab. VI, fig. 13, *a*—*b*.

FUSUS ALTUS. *S. Wood*. Catalogue 1842.

MUREX PULLUS (?). *Woodward*. Geol. of Norf. t. 3, fig. 27, 1833.

Tr. Testá turritá, altá, subulatá, tenui; anfractibus 7—8, convexis, longitudinaliter obsolete costatis, et transversim striatis; aperturá ovatá; labro tenui, intus lævigato; canali brevi vix recurvá.

Shell fusiform and turreted; very thin, with an obtuse apex; volutions 7—8, convex, with obsolete longitudinal costæ, and the remains of transverse striæ; aperture ovate, terminating in an open and short canal, slightly inflected; outer lip faintly sinuated.

Axis, nearly 2 inches.

Locality.　Red Crag, Butley, near Orford.

About a dozen specimens were found by myself many years since at a locality where a section of the Crag is not at present visible.　The specimens are evidently in an altered condition, like the attenuated specimens of *Buc. undatum*, var. *tenerum*, many of which were found with them at the same time.　The cause which reduced the substance of the one no doubt acted in the same way upon the other.　The canal is shorter and more open than in *T. gracile*, and it is a little imbricated behind the left lip by the reflected portions of the canal, like *Buc. undatum;* there were evidently longitudinal costæ upon the shell, vestiges of which are remaining upon some specimens on the upper volutions; and traces of transverse striæ may occasionally be seen, but the shells are nearly smooth.   It appears to be intermediate in form between *Strombus Norvegicus,* Chemn., and *Buc. fusiforme,* Brod. (Zool. Jour., 1829, t. 3, f. 3), and it may be said somewhat to resemble *F. Koninckii,* Nyst (pl. 40, f. 4), but that shell has a smaller and narrower canal, and a more acuminated apex. Should this species hereafter prove to be the same as Woodward's shell, my name must of course be given up.

5. TROPHON SCALARIFORME. *Gould.* Tab. VI, fig. 7, *a—c.*

> FUSUS SCALARIFORMIS. *Gould.* Rep. upon the Invert. of Massachus. p. 288, fig. 203, 1841.
> MUREX BAMFIUS (?). *Don.* Brit. Shells, pl. 169, fig. 1, 1799.
> —   PERUVIANUS. *J. Sow.* Min. Conch. t. 434, fig. 1, 1823.
> FUSUS LAMELLOSUS. *Gray.* Zool. of Beechey's Voy. pl. 36, fig. 13.
> TRITONIUM CLATHRATUM. *Lovén.* Ind. Moll. Scand. p. 12, 1846.
> FUSUS SCALARIFORMIS. *S. Wood.* Catalogue 1842.

Tr. Testá fusiformi, ventricosá, costellatá; utrinque attenuatá costis 15—20, *sublamellosis; anfractibus convexis, lævigatis* (?), *aperturá ovatá; canali elongatá recurvá.*

Shell fusiform and ventricose, attenuated at both extremities; volutions convex, smooth (?), suture deep; longitudinally costated, costæ or varices from 15—20, sublamellated, aperture with an elongated and recurved canal.

Axis, 1¼ inch.

Locality. Red Crag, Sutton and Bawdsey.

Mam. Crag, Bridlington.   Recent, North Seas and Massachusetts Bay.

This species is not very abundant in the Red Crag. It appears to correspond precisely with the shell brought from the North Seas, and it is identical with that found in the beds of the Clyde. The young of this species appears to differ from *Murex Bamfius,* Mont., in having a less number of reflected costæ, as well as in its projecting processes.   I have not seen the form of the true *Bamfius,* from the Crag. M. Lovén has united the two species in his Synopsis, under the name of *clathratum* (*Murex clathratus,* Linn.).   Our specimens from the Crag have been more or less rubbed, so that the transverse striæ, if they ever possessed them, are no longer visible. Among Mr. Bean's specimens from Bridlington, there is one with the name of *Bamfius* attached to it, but it appears to me to be only the young of this species.

The shell named *Fusus scalariformis,* figured by Nyst (pl. 40, fig. 5), is a different species.

6. TROPHON COSTIFERUM. *S. Wood.* Tab. VI, fig. 9, *a—b.*

> FUSUS RUGOSUS. *J. Sow.* Min. Conch. t. 34, 39, 199. 1817.
> MUREX RUGOSUS. *Parkinson.* Org. Rem. vol. iii, t. 5, fig. 16, 1811.
> FUSUS COSTATUS. *J. Sow.* Min. Conch. Syst. Ind. 1835.
> —   *S. Wood.* Catalogue 1842.

Tr. Testá elongatá, fusiformi, crassá ; anfractibus septem, convexis ; suturis profundis ; longitudinaliter costatis, costis 10—14 obtusis, transversim striatis ; aperturá ovatá ; labro intus incrassato, lævigato ; canali brevi subrecurvá.

Shell elongato-fusiform, thick, and strong, with seven convex volutions ; longitudinally costated, with 10—14 obtuse costæ ; transversely striated ; aperture ovate ; outer lip thickened, without denticulations ; canal short, slightly recurved.

Axis, 2 inches.

Locality.   Cor. Crag, Gedgrave.

Red Crag, Walton Naze, Sutton.

This is a very rare species in the Coralline beds, but abundant in the Red Crag, more especially at Walton, where it is found in good preservation.   Var. *b* has about thirteen costæ on the last volution.   It has fewer in its young state, and seldom has them in a continued line.   In var. *a*, the costæ do not appear upon the body whorl ; the apex tapers to a fine point, and is smooth.   The name *costatus* being preoccupied, I have proposed the above in the place of it.

7. **TROPHON ALVEOLATUM.**   *J. Sow.* Tab. VI, fig. 8, *a—b.*
>   FUSUS ALVEOLATUS.   *J. Sow.* Min. Conch. t. 525, 1826.
>   —   *S. Wood.* Catalogue 1842.
>   —   *Nyst.* Coq. foss. de Belg. pl. 39, fig. 21, 1844.

Tr.   Testá turritá, elongatá, subulatá, fusiformi ; anfractibus convexis, supernè subangulatis; longitudinaliter costellatis, alveolatis ; costellis transversis tuberculosis ; aperturá subovatá ; canali breviusculá, subrectá.

Shell turreted, elongate, fusiform; volutions convex, with two elevated, transverse, tubercular ridges ; upper part of volution subangulated ; longitudinally costellated, decussated, and alveolated; aperture ovate, with a moderate-sized and nearly straight canal.

Axis, 1¾ inch.

Locality.   Cor. Crag, Gedgrave.

Red Crag, Sutton and Brightwell.

This species was, until lately, rare, but numerous specimens have been recently obtained from a newly-opened part of the Coralline Crag at Gedgrave.   There are four transverse ridges upon the body of the shell, the upper one of which is the largest, giving a subcarinated form to the volution ; the two lower ridges are covered by the succeeding whorl, leaving only two visible upon the spire.   The left lip is very thin in all my specimens; and not so distinct as it is represented in the Belgian shell.

8. **TROPHON CONSOCIALE.**   *S. Wood.* Tab. VI, fig. 11, *a—b.*

Tr. Testa conico-turriculatá, fusiformi, spirá elevatá; anfractibus septem convexis, tumidis, subcarinatis, decussatis, supernè planatis ; costis transversis tuberculosis ; aperturá ovatá ; columellá recurvá ; canali breviusculá.

Shell turreted and fusiform, with an elevated conical spire ; volutions seven, convex, tumid, covered with four or five elevated and tuberculated transverse ridges ; longitudinally decussated, aperture ovate; canal moderate in size, and open.

Axis, 1¾ inch.

Locality. Cor. Crag. Ramsholt and Gedgrave.

Red Crag. Sutton and Newbourne.

This shell has been separated from the preceding species for the following reasons: the volutions are more tumid, and it has three transverse ridges upon the whorls of the spire, while in *Tr. alveolatum* there are but two, and the canal in the latter is more recurved. Numerous specimens of this and of the preceding species have been obtained within the last few years from Gedgrave, all of which appear to preserve the distinctions I have described without any intermediate or connecting forms.

9. TROPHON IMPERSPICUUM. *S. Wood.* Tab. VI, fig. 12.

Tr. Testá elongato-fusiformi, angustá, turritá ; spirá elevatá ; anfractibus octo, convexis ; suturis profundis, longitudinaliter costulatis, striis transversis decussatis ; ultimo anfractu obsoletè costato ; aperturá ovatá ; canali elongatá, subrectá.

Shell elongate, fusiform, turreted, and subulate, with an elevated spire, and acuminated apex; volutions eight, longitudinally costellated; costæ small and numerous, obsolete upon the body whorl, coarsely striated transversely; aperture ovate, with an elongate, open, and nearly straight canal.

Axis, 1¼ inch.

Locality. Cor. Crag, Sudbourn.

I have as yet obtained but one specimen of this elegant shell, which somewhat resembles *Fusus Deshayesii,* Nyst (Coq. foss. de Belg. p. 502, pl. 40, fig. 3) ; but it appears to differ specifically in being more elongate, and in having a greater number of costæ. The outer lip is a little broken, but in other respects the specimen is in very good condition. The species must, however, be considered doubtful, and the present name only as provisional, until more and better specimens be obtained.

10. TROPHON MURICATUM. *Mont.* Tab. VI, fig. 5.

MUREX MURICATUS. *Mont.* Test. Brit. p. 262, pl. 9, fig. 2, 1803.
FUSUS ECHINATUS. *J. Sow.* Min. Conch. t. 199, fig. 4, 1820.
— *Phil.* En. Moll. Sic. vol. i, p. 206, t. 11, fig. 10, 1836.
FUSUS MURICATUS. *Thorpe.* Brit. Mar. Conch. p. 206, fig. 97, 1844.

Tr. Testá turritá, muricatá, fusiformi ; anfractibus sex, convexis ; longitudinaliter costellatis, transversim striatis, decussatis, subimbricatis ; aperturá ovatá, labro intus denticulato ; canali brevi recurvá.

Shell fusiform, turreted, and echinated; whorls six, convex, with deep suture, longitudinally costated, and coarsely striated transversely; aperture ovate, with a short and recurved canal; outer lip dentated within.

Axis, ⅜ of an inch.

Locality. Cor. Crag, Sutton.

Red Crag, Walton-on-the-Naze.   Recent, Britain.

An abundant shell in the Red Crag. When I compiled my Catalogue it was considered as distinct from the recent species of Montague, in consequence of its having the canal shorter and more recurved. I have, however, since seen recent specimens, corresponding in that character with our shell, in the cabinet of Mr. Hanley, and in all other respects it perfectly resembles the recent form.

11. TROPHON GRACILIUS. *S. Wood.* Tab. VI, fig. 14.

FUSUS GRACILIOR. *S. Wood.* Catalogue 1842.

Tr. Testá elongato-fusiformi, gracili, fragili ; spirá elevatá, subulatá; lineis elevatis transversis cinctá ; longitudinaliter obsoletè costatá ; anfractibus quinque, convexis ; aperturá angustatá ; canali longiusculá.

Shell elongato-fusiform, slender, and fragile, with an elevated and tapering spire ; volutions five, convex ; obsoletely costated longitudinally, with elevated transverse striæ ; aperture elongate ; canal wide, and slightly recurved.

Axis, ⅜ of an inch.

Locality. Cor. Crag, Sutton.

But a few specimens of this species have come into my possession, and those, unfortunately, are not in very good condition ; it may be considered, therefore, at present a doubtful species. Traces of costæ are visible only on the spire, while the body whorl appears to be without them. It is a more slender shell than *Tr. imperspicuum* (fig. 12), with fewer costæ, and the volutions more rounded ; while the transverse striæ are fewer and more prominent. It has been figured more particularly for the purpose of calling the attention of collectors to its elegant form.

12. TROPHON PAULULUM. *S. Wood.* Tab. VI, fig. 6.

FUSUS PAULULUS. *S. Wood.* Catalogue 1842.

Tr. Testá pusillá fusiformi; apice acuto ; anfractibus quinque convexis, ultimo cingulis tribus elevatis ornatá ; interstitiis cancellatis ; aperturá ovatá ; canali longiusculá.

Shell very minute and fusiform ; spire elevated, and apex acute; whorls convex ; the last one covered with three elevated transverse ridges, and having the intervening spaces finely cancellated ; aperture ovate, with a canal moderately elongated and open.

Axis, ⅛ of an inch.

Locality. Cor. Crag, Sutton.

I have but one specimen of this pretty little shell, which may possibly be the fry of some larger species, though I have not any that show such markings upon the upper volutions. The upper whorls are worn smooth, and it is only on the body of the shell that this cancellated ornament can be seen. It resembles in shape a small shell, figured by Lea (Contributions to Geology, pl. 5, f. 155), but that shell does not appear to be ornamented like our species. The cancellated exterior is scarcely shown enough in the engraving.

FUSUS,* *Lam.* 1801.

FUSUS INTORTUS (?). *Lam.*

—   *S. Wood.* Catalogue 1842.

A worn and mutilated specimen of a shell which bears a great resemblance to the above Eocene species. It has been for some years in my cabinet, and was found by myself in the Red Crag of Sutton. It is possible that it may be a London clay specimen washed out of that formation.

FUSUS PORRECTUS (?), *Brander.*

FUSUS PORRECTUS. *S. Wood.* Catalogue 1842.

MUREX PORRECTUS. *Brander.* Foss. Hant. pl. 2, fig. 36.

A mutilated specimen also of what appears to be the well-known Barton species. It was found by myself in the Red Crag of Sutton, but it is not in a condition to be fairly identified. This shell and *F. intortus* are deeply stained with the ferruginous tinge of the Red Crag, and were probably both introduced into this formation from the London clay.

PLEUROTOMA,† *Lam.* 1801.

TURRICULA. *Schum.* 1817.

TOMELLA. *Sw.* 1840.

Gen. Char. Shell fusiform and turriculate, often thick and strong, generally sulcated or striated transversely; rarely smooth, sometimes nodose or tuberculated, with an ovate aperture, terminating inferiorly in a canal more or less elongated; outer lip sharp and thin, furnished with a slit or sinus below the suture; columella smooth, nearly straight, and in the recent state it has an acuminated operculum, with the nucleus at the sharper and lower extremity.

This genus comprises a large number of species, the greater part of which are natives of tropical or subtropical regions. None in which the sinus is immediately in the side have hitherto been found in a living state in a latitude so high as that of the British Channel. To this character it is here intended to restrict the genus. In the Eocene formations this genus is exceedingly abundant, both in species and in individuals, as if the climatal conditions of the seas of that period were more immediately favorable to its development.

* Etym. *Fusus,* a spindle.
† Etym. Πλευρά, the side, and Τομή, an incision.

1. PLEUROTOMA INTORTA. *Broc.* Tab. VI, fig. 4, *a, b.*

MUREX INTORTUS. *Broc.* Coq. foss. Subapenn. p. 427, t. 8, fig. 17, 1814.

PLEUROTOMA INTORTA. *Grat.* Tab. Coq. foss. Env. de Dax, p. 323, 1838.

— *S. Wood.* Catalogue 1842.

— *Morris.* Catal. of Brit. Foss. p. 157, 1843.

— *Nyst.* Coq. foss. de Belg. p. 509, pl. 41, fig. 2, 1844.

Pl. Testá elongatá, turritá, subfusiformi; longitudinaliter costatá; transversim striatá vel sulcatá; anfractibus supernè excavatis; cariná nodosá; aperturá ovato-angustá; labro infernè incrassato; canali brevissimá.

Shell elongato-fusiform, turreted, longitudinally costated, and transversely sulcated; volutions carinated and nodulous, concave on the upper part; aperture elongato-ovate, with a very short canal; outer lip thickened below.

Axis, 2¼ inches.

Locality. Red Crag, Butley and Sutton.

I have but two specimens of this species, one of which was found by my friend the late Rev. G. R. Leathes. Both specimens are somewhat rubbed, but sufficiently perfect to be identified with Brocchi's figure and the Belgian shell. The apex is much worn, so that the number of volutions cannot be correctly ascertained. The sinus appears to have been a little below the angle or keel of the volution; and longitudinal lines, or lines of growth, are slightly visible.

2. PLEUROTOMA TURRICULA. *Broc.* Tab. VI, fig. 1, *a—b.*

MUREX TURRICULA. *Brocchi.* p. 435, t. 9, fig. 20, 1815.

PLEUROTOMA TURRICULA. *Bronn.* Ital. tert. geb. p. 46, 1831.

— *Philip.* En. Moll. Sic. p. 199, 1836.

— *Nyst.* Coq. foss. de Belg. p. 520, pl. 41, fig. 5, 1844.

Pl. Testá elongato-fusiformi; anfractibus subcarinatis; cingulis transversalibus tribus remotis, elevatis, obtusis, interdum subcrenulatis; interstitiis transversim et longitudinaliter tenuissimè striatis; aperturá oblongá, canali angustá rectiusculá.

Shell elongato-fusiform, whorls subcarinated, with three remote, elevated, and obtuse, transverse bands, which are sometimes subcrenulated; finely striated transversely; lines of growth visible; aperture elongated, canal produced.

Axis, 1½ inch.

Locality. Red Crag, Sutton and Bawdsey.

This is also a rare species. My cabinet contains about a dozen specimens, but none in very good condition. I have in consequence been obliged to borrow parts of the description of the shell from M. Nyst, the subcrenulations and fine striæ not being visible in my specimens, although, from their form and general character, I have no doubt of the identity. The sinus is exactly at the angular part of the volution, above which point it is rather concave, with an elevated band near the suture. The longitudinal striæ I presume are the fine lines of growth, more especially visible where the shell has lost part of its outer coating. I have figured two individuals, both of which are rubbed and worn.

3. PLEUROTOMA CARINATA. *Biv.* Tab. VI, fig. 2, *a, b.*

PLEUROTOMA CARINATUM. *Bivonæ* in *Philip.* Enum. Moll. Sic. vol. ii, p. 176, t. 26, fig. 19, 1844.

Pl. Testá fusiformi, turritá; anfractibus septem, lævissimis, subangulatis, carinatis, supernè concavis, infra convexis; aperturá ovatá; canali elongatá; labro acuto.

Shell turriculate and fusiform; volutions smooth, with a prominent keel about the middle, concave above the keel, convex below it; aperture ovate; canal elongated, and rather broad.

Axis, ¾ of an inch.

Locality. Cor. Crag, Gedgrave.

Red Crag, Sutton.

I have but one specimen from each formation. These shells have about eight volutions, with the keel a little above the centre of the whorl. Dr. Philippi describes his shell as having a prominent and elevated keel. In the Crag specimens it is only sharply angulated, being, perhaps, worn at that part. The position of the sinus is not distinctly defined in either of my specimens. Probably it was a little above the keel.

4. PLEUROTOMA SEMICOLON (?). *J. Sow.* Tab. V, fig. 3, *a—b.*

PLEUROTOMA SEMICOLON. *J. Sow.* Min. Conch. t. 146, fig. 6, 1816.

— COMMA. - - - - t. 146, fig. 5.

Pl. Testá elongato-fusiformi, turritá; anfractibus numerosis convexis, supernè concavis, medio nodulosis; transversim striatis vel cingulatis, prope suturam granulatis; aperturá elongato-ovatá; canali angustá.

Shell elongate and fusiform, variable, whorls convex, covered with striæ or bands; spire crenulated, aperture elongate, outer lip curved; canal rather variable in length.

Axis, ¾ of an inch.

Locality. Cor. Crag, Gedgrave.

Red Crag, Sutton.

I have but a few specimens of this species in good preservation. It appears, however to correspond with one of the varieties of that variable and well-known shell from Barton and Bracklesham, where it is abundant. This is one of the few Eocene shells that may be considered as having lived on to the Crag period. The sinus of our shell is situate in the upper and most projecting part of the volution, at which place there are two rows of nodules; these are carried up the spire like the marks of punctuation which suggested the name to Mr. Sowerby. The Crag species appears a variety of *P. semicolon;* there is one ridge near the suture, above the concave part of the volution, and the lip curves elegantly from the sinus. The only difference I can observe in the shells from the two different formations is in the arrangement of the transverse striæ, which are broader and more regular in the Crag shell than they are upon the Barton shells, but living under altered conditions would in all probability modify these appearances.

5. PLEUROTOMA PORRECTA. *S. Wood.* Tab. VII, fig. 1, *a—b.*

PLEUROTOMA PORRECTA. *S. Wood.* Catalogue 1842.

Pl. Testá porrectá, elongato-fusiformi; spirá turritá, apice acuto; anfractibus 9—10, *parum convexis, subangulatis, transversim tenuissimè striatis; in medio nodulosis; canali longuisculá rectá.*

Shell elongato-fusiform, with an elevated spire and acute apex; volutions 9—10, slightly convex, finely striated transversely, and furnished in the middle of each whorl with one row of obtuse, ovate nodules; aperture ovate; canal elongated.

Axis, $1\frac{1}{8}$ of an inch.

Locality. Cor. Crag, Gedgrave.

When my Catalogue was drawn up I possessed but one mutilated specimen of this species. I have, however, obtained another since that time, and with three more belonging to H. Daniel, Esq. I am enabled to define its characters. The length of the aperture, including the canal, exceeds two fifths of the entire length of the shell, but in many species that character is liable to great variation, and is not much to be depended upon. The nodules are obtuse, oblique, and placed in the centre of the whorls; they are produced by the reflected edge of the lower part of the sinus in the outer lip, which being thickened externally at different periods, leaves permanent nodules which ornament the exterior. They are about ten in the penultimate whorl, but become nearly obsolete on the older part of the shell. It is covered with very fine impressed striæ, the spaces between them being broad, flat, smooth, and glossy, differing in that respect from the rough, elevated striæ upon *Pl. nodularia,* Desh. (Coq. foss. de Par. p. 493, t. 66, f. 23—25). In other respects it much resembles that shell. It might perhaps be considered as *Pl. noduliferum,* Phil. (En. Moll. Sic. vol. ii, p. 173, t. 26, f. 16), only that he has described that shell as " lævissima," and his figure does not appear quite so tapering as the one from the Crag, and it has also a shorter canal. When, hereafter, the shells are compared, they may prove the same, but not knowing the Sicilian species I have left the Crag shell with its provisional name. A shell in Mr. Lyell's cabinet from Touraine is probably identical with the Crag specimens.

6. PLEUROTOMA NODULOSA (?). *Desh.*

PLEUROTOMA NODULOSA (?). *Desh.* Coq. foss. des Env. de Paris, pl. 65, fig. 11-14.

A few fragments resembling this Eocene species are in my cabinet, but they are too imperfect for figuring or fair description. My best specimens appear to have had seven convex volutions, covered with obtuse nodules, about a dozen in each volution, with half a dozen elevated transverse lines.   At the upper part, near the suture, there is a row of small denticulations.   The first two volutions are smooth, with a very obtuse apex, in which character it disagrees with the French fossil, where it is sharp and pointed.

CLAVATULA,* *Lam.* 1801.
MANGELIA. *Leach. Risso.* 1826.
DEFRANCIA. *Miller.* 1826.
MANGILIA. *Lovén.* 1846.

Gen. Char. Shell fusiform, turreted, rarely of any great size, generally ribbed in a longitudinal direction, and transversely striated; aperture ovate, terminating in a somewhat short, open, canal; outer lip sharp, sometimes denticulated within, with a small sinus at the suture.

This name was proposed by Lamarck for a division of the Pleurotomæ, to contain those species with a small and shallow sinus, which he afterwards, in his 'Hist. des An. sans Vert.,' united to Pleurotoma. The shells here intended to be grouped together are those in which the "side slit" is represented by a somewhat obscure sinus, and that, instead of being in the outer lip, is situated at the upper extremity of the aperture, at or near the junction of the body whorl. The name Mangelia was given in MS. by Dr. Leach to some shells of this character in the British Museum; and this name was adopted by Risso, in his 'Hist of the Shells of Merid. Europe.' Messrs. Reeve and Sowerby have recently employed it for the same shells; Lamarck's name is, however, of prior date.

Many recent species of this genus are found within the tropics, others have been obtained in the Arctic seas,

1. CLAVATULA LINEARIS. *Mont.* Tab. VII, fig. 2, 2 *a.*
            MUREX LINEARIS. *Mont.* Test. Brit. p. 261, t. 9, fig. 4, 1803.
            — ELEGANS. *Don.* Brit. Shells, t. 179, fig. 3.
            FUSUS LINEARIS. *Flem.* Brit. An. p. 350, 1828.
            PLEUROTOMA LINEARIS. *Thorpe.* Brit. Mar. An. p. 197, 1844.
            —   *S. Wood.* Catalogue 1842.
            MANGELIA LINEARIS. *Leach.*
            DEFRANCIA LINEARIS. *Lovén.* Ind. Moll. Scand. p. 13, 1846.

Cl. Testá turriculá, fusiformi, subventricosá, spirá elevatá, acuminatá; anfractibus convexis, longitudinaliter costatis; transversim lineatis, lineis elevatis, striatis, labro incrassato, intus lævi; canali brevi.

Shell fusiform and turriculate, with an elevated and acute spire; whorls convex, longitudinally costated, and sharply striated transversely; lip thickened on the outside, smooth within; canal short, rather broad.

* Etym. (?) *Clavatula,* the diminutive of *clava,* a club.

Axis, ⅜ of an inch.

Locality. Cor. Crag, Sutton.

Red Crag, Sutton. Recent, British Seas.

This elegant little shell has about seven whorls, which are ornamented with a dozen costæ in the last volution ; these are crossed by sharp elevated lines, which run along the volutions ; they are rather broader over the ribs, and very sharp between them, and are from six to seven in number. The outer lip is sharp, strengthened on the exterior by a thickened obtuse varix or rib ; canal short, open, and slightly inflected. It is by no means abundant in either the Red or Coralline Crag formations.

2. CLAVATULA PHILBERTI. *Mich.* Tab. VII, fig. 5, 5 *a.*

PLEUROTOMA PHILBERTI. *Michaud.* Phil. En. Moll. Sic. tom. ii, p. 165, 1844.

— VARIEGATU *Phil.* En. Moll. Sic. vol. i, p. 197, tab. 11, fig. 14, 1836.

— — *S. Wood.* Catalogue 1842.

Cl. Testá turritá, fusiformi ; anfractibus convexis, costis longitudinalibus lineisque transversis elevatis, clathratis ; labro incrassato, intus dentato ; canali breviusculá.

Shell turreted and fusiform, with about six convex volutions, longitudinally costated, and decussated by coarse, elevated, transverse striæ ; outer lip thickened within and dentated.

Axis, ⅜ of an inch.

Locality. Cor. Crag, Sutton. Recent, Mediterranean.

I have one tolerably good specimen of this species, with a few imperfect ones, and have very little doubt of the identification. The striæ are sharp and elevated, like those upon the preceding species, but coarser, making the costæ rather tuberculate ; the lip is thickened within, somewhat contracting the aperture, and strongly dentated, with seven or eight elevated teeth, and having a moderately sized sinus immediately at the suture. The figure is scarcely elongated enough.

3. CLAVATULA CASTANEA (?). *Brown.* Tab. VII, fig. 3, 3 *a.*

FUSUS CASTANEUS. *Brown.* Illust. Brit. Conch. pl. 48, fig. 43-44, 1827.

Cl. Testá pusillá, ovato-fusiformi, anfractibus quatuor valdè convexis, tumidis, prope suturam depressis ; longitudinaliter obtusè costatis, costis 11—12, transversim striatis, striis rugosis, confertis ; aperturá ovatá ; canali brevi.

Shell small, ovato-fusiform, with four convex and tumid volutions ; depressed near the suture, with 11—12 obtuse costæ, numerously and coarsely striated transversely ; striæ carried over the ribs ; aperture ovate ; canal short.

Axis, ⅕ of an inch.

Locality. Cor. Crag, Sutton. ? Recent, British Seas.

I have but two specimens of this species, and these appear to resemble the figure by Brown, referred to above. My specimens are thickly and rather coarsely striated,

8

the striæ being carried over the ribs; this character is not represented in Brown's figure.   The sinus in my specimens is at the suture, but not very distinct, and there is a slight depression on the upper part of the whorl.   It differs from *Cl. perpulchra* in being shorter, in having the volutions more tumid, and in its coarser and more numerous striæ.   It is a much less elegant shell.

4. CLAVATULA PERPULCHRA.   *S. Wood.*   Tab. VII, fig. 4, 4 *a*.

Cl. Testá minutá, fusiformi, turriculá; spirá elevatá; anfractibus convexis; longitudinaliter costatis, transversim striatis; striis 10—11 acutis, elevatis; labro extus incrassato, intus lævi; canali brevi; sinu apud suturam profundo.

Shell small, turriculate, and fusiform; with an elevated spire and convex volutions; longitudinally costated; costæ 11—12 in the last volution; transversely striated; striæ sharp and elevated; outer lip curved, and thickened on the outside, with a large and deep sinus at the suture; canal short, and very open.

Axis, ¼ of an inch.

Locality.   Cor. Crag, Sutton.

This species much resembles in form and size *Cl. linearis*, but that shell has fewer striæ, and the ribs are narrower; in this the ribs are broad, touching each other at the base, leaving no space between them; whereas, in *Cl. linearis*, they stand apart, and the transverse striæ or ridges are fewer and sharper, with the outer lip less curved, more contracted at the lower part, and having the sinus not quite so deep as in this species.

Imperfect specimens of this shell are by no means rare at the above locality.   It was considered in my Catalogue only as a variety of *Cl linearis*, but, for the above reasons, they are now separated.

5. CLAVATULA COSTATA (?).   *Da Costa.*   Tab. VII, fig. 6 *a*, 6 *b*.

BUCCINUM COSTATUM.   *Da Costa.* Hist. Nat. Test. Brit. p. 128, tab. 8, fig. 4, 1778.
MUREX COSTATUS.   *Mont.* Test. Brit. p. 265, 1803.
FUSUS COSTATUS.   *Thorpe.* Mar. Brit. Conch. p. 202, 1844.
PLEUROTOMA MITRULA (?).   *Nyst.* Coq. foss. de Belg. p. 528, pl. 44, fig. 3, 1844.

Cl. Testá turritá, subulatá, fusiformi; anfractibus 6—7 convexiusculis, costatis; costis circa septem, obliquis, obtusis, elevatis; interstitiis transversim tenuissimè striatis; labro acuto, extus marginato, intus incrassato; canali brevi.

Shell turriculate, tapering, subfusiform, with 6—7 slightly convex volutions; longitudinally costated; costæ seven, oblique, elevated, and obtuse, transversely striated, with an acute and curved outer lip, thickened within, particularly near the sinus; and marginated on the outside behind the lip.

Axis, ⅝ of an inch nearly.

Locality.   Cor. Crag, Sutton.

Red Crag, Sutton and Walton.   ? Recent, British Seas.

This species is not rare in the Red Crag, but the specimens are generally more or less rubbed. In those which are best preserved, fine impressed striæ may be detected between the ribs. It differs from *Cl. mitrula* in having fewer costæ, and a more elevated spire; and the ribs in this are more oblique, sloping to the left, and preserving generally a continuous oblique line of ridge, interrupted only by the suture; the spaces between them are rather wider than the ribs, and the first two volutions are destitute of costæ, but are cancellated; the aperture varies from one third to two fifths of the length of the axis. The figure does not well represent the ribs, as they slope a little more towards the upper part, and incline backwards. The outer lip is sharp, with a thickened obtuse varix on the exterior, and it has also a thickening upon the inside, particularly near the sinus, which is rather broad. The figure by Nyst probably represents the same species, though my specimens are all more slender and attenuated than his figure. It is, however, a doubtful identification, and he has described that shell as smooth; my Crag specimens, when perfect, are transversely striated.

6. CLAVATULA MITRULA. *J. Sow.* Tab. VII, fig. 7, 8 *a*.

        BUCCINUM MITRULA. *J. Sow.* Min. Conch. t. 375, fig. 3, 1822.

        PLEUROTOMA MITRULA. *S. Wood.* Catalogue 1842.

Cl. Testá turritá, elongatá, subfusiformi, apice acuto; anfractibus septem convexiusculis, plicatis, plicis circa decem obliquis, sub lente tenuissimè striatis; striis transversis impressis; aperturá lanceolato-ovatá.

Shell elongato-turriculate, subfusiform, costated; whorls 7, slightly convex, ornamented with about ten oblique, obtuse, rounded ribs, and finely striated transversely, with a sharp and curved outer lip; sinus a little below the suture.

Axis, $\frac{5}{8}$ of an inch.

Locality. Cor. Crag, Sutton.

        Red Crag, Sutton and Walton Naze.

This is an abundant species in the Red Crag. Good specimens are covered with transverse impressed striæ. The ribs are very slightly curved, and are rather more prominent on the upper part; they are as wide as the spaces between them, and rarely in a continuous line; the apex is sharp, and the first two volutions are without the large obtuse ribs, but are cancellated. The outer lip is slightly curved and sharp; it is marginated on the outside, and thickened within, particularly below the sinus, and has a deep, distinct suture. The lower part of the outer lip projects slightly, with a wide and rather short canal. Aperture about two fifths the length of axis. It resembles in form *Pl. multilineolatum*, Phil. (t. 26, f. 1), but the whorls are more convex.

7. CLAVATULA BRACHYSTOMA. *Phil.* Tab. VII, fig. 8, 8 *a*.

    PLEUROTOMA BRACHYSTOMA. *Phil.* En. Moll. Sic. p. 169, t. 26, fig. 10, 1844.

Cl. Testá turritá, subulatá, fusiformi; anfractibus convexis, supernè angulatis; costatis, costis longitudinalibus novem, ad suturam evanescentibus; striis elevatis transversis, aperturá ovatá, canali brevi; labro acuto, extus incrassato.

Shell turreted and tapering; apex rather obtuse; whorls convex, slightly angulated at the upper part; costated and striated; costæ 9, aperture ovate, with a short canal; outer lip sharp, thickened externally.

Axis, $\frac{3}{8}$ of an inch.

Locality. Cor. Crag, Sutton.   Recent, Mediterranean.

This species is by no means rare in the Coralline Crag. It agrees with a recent specimen of *brachystoma* (obligingly lent me for comparison by Mr. Hanley, and received by that gentleman from M. Philippi), except that the striæ appear a little finer. The upper three volutions in the specimen from the Crag are cancellated, and on the third volution there are at least a dozen ribs, whereas on those succeeding it there are never more than eight or nine. The sinus in our shell is in the angulated part of the volution, giving a slight depression at the upper part of the whorl.

8. CLAVATULA NEBULA. *Mont.* Tab. VII, fig. 10.

    MUREX NEBULA. *Mont.* Test. Brit. p. 267, t. 15, fig. 6, 1803.
    PLEUROTOMA BERTRANDI. *Phil.* En. Moll. Sic. vol. i, p. 198, t. 11, fig. 20, 1836.
    FUSUS (?) NEBULA. *S. Wood.* Catalogue 1842.
    FUSUS NEBULA. *Thorpe.* Brit. Mar. Conch. p. 203, fig. 93, 1844.
    PLEUROTOMA GINNANNIANUM. *Phil.* En. Moll. Sic. vol. ii, p. 168, tab. 26, fig. 6, 1844.
    MANGILIA NEBULA. *Lovén.* Ind. Moll. Scand. p. 13, 1846.

Cl. Testá turritá, elongatá, subulatá; anfractibus 7—8, convexis, longitudinaliter costulatis, transversim striatis; canali breviusculá rectá; labro haud incrassato.

Shell elongate, tapering, with an acuminated spire; volutions 7—8, convex; suture deep, spirally striated, and longitudinally costated; costæ 10—11; aperture ovate; outer lip sharp, with a small sinus a little below the suture; canal very short and open.

Axis, $\frac{7}{8}$ of an inch.

Locality. Red Crag, Sutton.   Recent, Mediterranean and British Seas.

I have as yet seen this species from one locality only, where it is not very rare. It appears to differ slightly from Philippi's figure and description. My specimens are most of them much rubbed, though in some the striæ are distinctly visible; Philippi's figures have fewer ribs. The proportions of the aperture are variable, though generally about one third the length of the shell. Lovén considers the *M. nebula* of British authors as identical with the *Ginnannianum* of Philippi, and I have therefore, upon his authority, retained the name originally given in my Catalogue, although the Crag specimens correspond better with the Mediterranean shell, and are larger than any recent British specimens I have seen.

9. **Clavatula cancellata.** *J. Sow.* Tab. VII, fig. 9.

    Fusus cancellatus. *J. Sow.* Min. Conch. t. 525, fig. 2, 1827.
    Pleurotoma cancellata. *S. Wood.* Catalogue 1842.

Cl. Testá turritá elongato-fusiformi; anfractibus 6—7, convexis; supernè planatis, depressis, longitudinaliter costatis et transversim striatis clathratis; papillis in angulis sectionem acutis; canali longiusculá, subrectá.

Shell slender, fusiform, and elongate; apex acute; spire elevated, with 6—7 convex volutions, covered with decussating ridges, which are sharp and elevated at the point of contact; aperture ovate; canal moderately long, very slightly recurved.

Axis, ⅞ of an inch.

Locality. Cor. Crag, Sutton.
    Red Crag, Walton Naze.   ? Recent, Mediterranean.

This elegant shell is at present rare from both formations. The sinus is rather broad, producing a flattened or smooth canal around the spire, immediately at the suture. The whorls are covered with longitudinal ribs, and with transverse ridges, elevated and spinous at the points of junction. The lower part of the aperture is slightly contracted, forming a distinct canal, which is open and very slightly recurved. This is probably only a variety of *Pleurot. reticulata*, Bronn., *Murex echinatus*, Broc. (Tab. 8, fig. 3), *Pleurot. Cordieri*, Payr.

10. **Clavatula concinnata.** *S. Wood.* Tab. VII, fig. 11 *a*, 11 *b*.

    Pleurotoma rufa (?). *S. Wood.* Catalogue 1842.

Cl. Testá turriculá, elevatá, fusiformi; anfractibus convexiusculis, prope suturam obsoletè angulatis; longitudinaliter obliquè plicatis, plicis 10—13; transversim striatis; aperturá elongato-ovatá; labro acuto; canali breviusculá.

Shell turriculate, elevated, and fusiform; with an obsolete angularity a little below the suture; longitudinally plicated; plicæ or ribs oblique; crossed and cancelled by coarse and elevated striæ, between which are finer lines; outer lip sharp; canal moderate.

Axis, 1 inch.

Locality. Cor. Crag, Sutton and Gedgrave.

This is not a very abundant shell, and is seldom perfect. The large figure is from a specimen belonging to Mr. Perry. The outer lip curves a little, giving an elegant obliquity to the folds, and the sinus is shallow a little below the suture. It is crossed by about six or seven elevated lines, which decussate the exterior, and the shell is covered between these with finer lines. It somewhat resembles the figure of *F. rufus*, Gould (Invert. Massach., p. 290, f. 192), but differs from the British shell *M. rufus*, Montague. It is, however, possible that it may be only an extreme variety of *C. turricula*. Fig. 11 *b* is a more elongate variety.

11. CLAVATULA LÆVIGATA (?). *Phil.* Tab. VII, fig. 12, 12 *a*.

PLEUROTOMA LÆVIGATUM. *Phil.* En. Moll. Sic. vol. i, t. 11, fig. 17.

Cl. Testá fusiformi, turritá, subulatá; transversim tenuissimè striatá, anfractibus plani-usculis, contiguis; aperturá longitudine totius testæ trientem vix æquante; canali brevissimá.

Shell turreted, tapering, and fusiform; with an ovate or elongated aperture; whorls slightly convex, finely striated transversely, more coarsely so at the base, with a short and wide canal, columella slightly incurved, outer lip sharp, and smooth within.

Axis, $\frac{9}{16}$ of an inch.

Locality.   Cor. Crag (?), Sutton.

Red Crag, Walton Naze.   Recent, Mediterranean.

A few specimens only of this species have come into my possession, and those unfortunately are not in good condition.   Mr. Bowerbank has also lent me two speci-mens from Walton Naze; but they are unfortunately in no better condition than my own. I feel, however, unwilling they should not be represented.   It much resembles *Pleu-rotoma lævigatum*, Phil. (vol. i, p. 199, t. 11, f. 17), and if it were not for his description, "*lævissima*," I should have little doubt of their identity; our shell is distinctly striated over the body of the whorls, and more coarsely so around the base.   The obsolete costæ are scarcely visible, from the bad condition of the specimens, and there is a visible depression at the upper part of the whorl, caused by a deep sinus a little below the suture.

12. CLAVATULA TURRICULA. *Mont.* Tab. VII, fig. 13 *a*, 13 *b*.

MUREX TURRICULA. *Mont.* Test. Brit. p. 262, t. 9, fig. 1, 1803.

—   ANGULATUS. *Don.* Brit. Shells, t. 156.

—   PUNCTATUS. *Woodward.* Geol. of Norf. t. 3, fig. 28, 1833.

PLEUROTOMA CLAVULA. *Dujard.* Mém. de la Soc. Géol. de France, 1837, tom. xi, pt. 2, p. 291.

DEFRANCIA NOBILIS. *Möller.* Ind. Moll. Grœnl. p. 12, 1842.

—   WOODIANA.   -   -   -   -   p. 13.

FUSUS PLICATILIS. *Bean.* MS.

Cl. Testá turriculá, ovato-fusiformi; anfractibus convexiusculis, supernè angulatis; longitudinaliter costulatis; transversim et rugosè striatis; apertura ovatá, canali brevi.

Shell turriculate, ovately-fusiform; whorls about eight, slightly convex, angular at the upper part; longitudinally costulated, with 14—16 slightly elevated ribs in the last volution, crossed by coarse transverse striæ; aperture subovate, ending in a short and rather wide canal.

Axis, $\frac{7}{8}$ of an inch.

Locality. Red Crag, Sutton and Walton.

Mam. Crag, Bramerton and Bridlington.   Recent, British and North Seas.

This is by no means a rare species in the Red Crag.   It is rather an aberrant form of this genus, having the sinus indistinctly marked by a slight indenture at the angle, and not immediately at the suture. The transverse striæ are continued over the ribs when the shell is perfect, but those prominent parts are often smooth. The costæ are in general vertical, though in some there is a slight curvature. The specimens from Bramerton appear punctated, from an alteration and erosion of the surface. A specimen from Bridlington, sent by Mr. Bean, with the name "*plicatilis*," appears to be a variety of this species, which is very variable in its proportionate dimensions as well in the fossil as in the recent specimens; several shells, considered as specifically distinct by Möller, are probably only varieties.

13. CLAVATULA TREVELLIANA. *Turt.* Tab. 7, fig. 14.

PLEUROTOMA TRAVELLIANA. *Turt.* Mag. Nat. Hist. p. 351, 1834.
—   *Thorpe.* Brit. Mar. Conch. p. 197, fig. 52, 1844.
—   *Macgyllivray.* Moll. Aberd. p. 172.
FUSUS (?) TURRICULA, var. *S. Wood.* Catalogue 1842.

Cl. Testá ovato-fusiformi, spirá subturritá; apice acuto; anfractibus 5—6, convexis, subinflatis, supernè angulatis; longitudinaliter costulatis, costulis confertis; transversim rugosè striatis, cancellatis; aperturá ovatá, canali brevissimá.

Shell ovato-fusiform, with a slightly elevated spire, and acute apex; whorls 5—6, convex and rather tumid, angulated at the upper part; longitudinally costulated with numerous rather oblique and slightly prominent ribs, and large transverse striæ decussating and cancellating the exterior.

Axis, ½ an inch.
Locality. Red Crag, Sutton.
         Mam. Crag, Bramerton.   Recent, British Seas.

This shell appears to correspond with the figure and description of what is considered a distinct species, and I have given it upon the authority of British conchologists, who may perhaps have better means of determining that point either from colour or other character. I have but two or three specimens; it may, however, be a variety of *Cl. turricula*, as it is only in the proportions that I can observe any difference between my fossil specimens of this species and those of *turricula*.

14. CLAVATULA BOOTHII. *Smith.* Tab. VII, fig. 16.

PLEUROTOMA PURPUREA. *Bast.* Mém. Géol. de Bord, pl. 3. fig. 10, 1825.
FUSUS BOOTHII. *Smith.* Wern. Mem. vol. viii, p. 51, pl. 1, fig. 1, 1838.

Cl. Testá fusiformi, turritá, longitudinaliter costatá; anfractibus convexis, transversim lineatis, lineis elevatis; labro obtuso, intus nitido; canali brevi subrectá.

Shell turreted and fusiform, longitudinally costated; whorls convex, slightly inflated, covered with elevated striæ; outer lip obtuse, smooth within; canal short, and slightly recurved.

Axis, ¾ of an inch.
Locality. Red Crag, Walton Naze.   Recent, British Seas.

Of this species I have but a few specimens, and those appear restricted to the Red Crag. It approaches in form *Pl. inflatum*, Phil. (En. Moll. Sic. vol. i, t. 11, f. 24), *Pl. Leufroyi* (vol. ii, p. 165), and upon comparison with specimens of that shell may possibly be found to be the same species. The Crag shell has 12—13 costæ; is rather more elongate, and has a longer canal; the striæ upon the exterior are alternately one large and one small. The outer lip is contracted at the lower part, forming a distinct but open canal, and the costæ are not carried below the body of the whorl. *Pleurotoma purpurea*, Basterot, is in all probability the same as our shell, which differs from *purpurea*, Montague, in being less distinctly cancellated, and in having larger and more elevated ribs; it is also a shorter shell.

15. CLAVATULA PLICIFERA. *S. Wood.* Tab. VII, fig. 15.

PLEUROTOMA PLICIFERA. *S. Wood.* Catalogue.

Cl. Testâ elongato-fusiformi, turritâ; anfractibus convexis, supernè depressis, longitudinaliter plicatis, plicis numerosis, subsinuosis; transversim striatis; labro acuto; canali longiusculâ.

Shell elongato-fusiform, turreted; whorls subangulated, with longitudinal subsinuated folds, or costulæ; transversely striated; aperture ovate; canal moderate.

Axis, $\frac{3}{4}$ of an inch.

Locality. Red Crag, Sutton.

But one specimen of this species is in my cabinet. It appears distinct from any species I have seen, in the peculiar curvature of the outer lip, which gives an elegant flexion to the folds or ribs. The sinus is rather below the suture, like that in *Cl. turricula,* but the costæ in that shell are much more erect. It is, however, not a well-determined species, and the above name may be considered as provisional until more perfect specimens are found. It may be here remarked, that many of the specimens belonging to this genus, found in the Crag, are far from being in good condition, and the identifications of some of the species are by no means satisfactory.

CANCELLARIA,* *Lam.* 1801.

PURPURA, (spec.) *Adanson.*
ADMETA. *Kröyer.*

Gen. Char. Shell ovate, fusiform or turriculate, costated; aperture subcanaliculated, sometimes entire; columella plicated, plicæ variable; outer lip generally sulcated within.

1. CANCELLARIA CORONATA. Tab. VII, fig. 18 *a*, 18 *b*.

CANCELLARIA CORONATA. *Scacchi.* In Phil. En. Moll. Sic. vol. ii, p. 177, t. 25, fig. 24, 1844.
— VARICOSA. *Phil.* En. Moll. Sic. vol. i, p. 201, 1836.
— — *Nyst.* Coq. foss. de Belg. p. 475, pl. 38, fig. 20.
— LÆVICOSTA. *S. Wood.* Catalogue 1842.

* Etym. *Cancelli,* lattices; from the exterior of the shell being covered with crossed lines, like latticework.

C. Testá turritá, elongatá, fusiformi, spirá elevatá; apice acuminato; longitudinaliter costatá; anfractibus supernè angulatis, supra planis; costis distantibus rotundatis, lævigatis, interstitiis obsoletè striatis; columellá biplicatá; labro intus sulcato.

Shell elongato-fusiform, turreted, with an elevated spire, longitudinally costated, obsoletely striated between the ribs; whorls subangulated above, with the costæ produced at the upper part, smooth and distant; columella with two folds, outer lip denticulated within; aperture ovate, slightly canaliculate.

Axis, 1¼ inch.

Locality. Red Crag, Walton Naze.

This species is very rare in my cabinet. It corresponds with the figure and description given by Nyst, at the above reference, but I cannot consider it *Vol. varicosa* of Brocchi, unless it be very much altered. It is, I imagine, the same as the one Dr. Philippi has figured in his second volume, although the shape of the outer lip in his figure is more expanded at the lower part, and his shell appears to have more and sharper costæ. In my specimens they are never more than ten in one volution, and the shell is not quite so much coronated, although the costæ are produced at the upper part, but they are more obtuse. The specimens are somewhat altered, as the striæ are nearly obsolete, but the ribs appear smooth. In my Catalogue it was considered as a new species, and the present identification rests upon Dr. Philippi's description, who considered it in his first vol. as *Vol. varicosa*, Broc., and says, p. 201, " labro sulcato;" in the second vol. p. 177, the character is given as " labro intus lævi." It may perhaps be so in its young state.

2. CANCELLARIA MITRÆFORMIS (?). *Broc.* Tab. VII, fig. 19, *a—b.*

     VOLUTA MITRÆFORMIS. *Broc.* Conch. foss. Subap. p. 645, t. 15, fig. 13, 1815.

     CANCELLARIA MITRÆFORMIS. *S. Wood.* Catalogue 1842.

C. Testá turritá, subfusiformi; longitudinaliter obsoletè costulatá; transversim striatá, columellá biplicatá, labro acuto, intus sulcato, canali brevissimá.

Shell turreted, elevated, subfusiform; with obsolete longitudinal costæ; transversely striated; columella biplicate; outer lip sharp, dentated within, with a subemarginate base.

Axis, ⅝ of an inch.

Locality. Cor. Crag, Gedgrave.

             Red Crag, Sutton.

This species is rare. The upper part of the volution is flattened, while it is convex on the lower half. The outer lip is sharp, somewhat expanded, thickened within, and has about nine or ten denticulations. The striæ upon the exterior are large and coarse, amounting almost to ridges, and the longitudinal costæ in some specimens are obsolete.

Our shell has not the ridge-like appearance below the suture so visibly shown in Brocchi's figure, nor are there any denticulations represented in his figure. The identification is therefore given with doubt.

9

3. CANCELLARIA COSTELLIFERA. *J. Sow.* Tab. VII, fig. 21.

MUREX COSTELLIFER. *J. Sow.* Min. Conch. t. 119, fig. 3, 1818.
CANCELLARIA COUTHOUYI. *Jay.* Cat. of his Cab. 1839.
— BUCCINOIDES. *Couthouy.* Bost. Journ. of Nat. Hist. vol. xi, p. 105, pl. 3, fig. 3.
— GRANULATA. *Nyst.* Coq. foss. de Belg. p. 479, pl. 39, fig. 14, 1844.
— VIRIDULA. *Lovén,* Ind. Moll. Scand. p. 15, 1846.

C. Testá ovato-fusiformi, ventricosá; spirá acuminatá; anfractibus convexis, longitudinaliter costellatis, transversim striatis; costellis 12—15, granulatis; in ultimo anfractu ad basim evanescentibus; aperturá ovatá; columellá triplicatá; basi acuminatá; canali brevissimá.

Shell ovato-fusiform, with an acuminated spire; whorls convex, obtusely costated longitudinally; costæ 12—15, generally obsolete upon the lower half of the body whorl, transversely striated; aperture acuminated at the base, with a very short canal; columella triplicate; outer lip sharp.

Axis, $\frac{5}{8}$ of an inch nearly.

Locality.   Cor. Crag, Sutton.
Red Crag, Sutton.
Mam. Crag, Bridlington.   Recent, Coast of Massachusetts.

This species first appears in the Coralline Crag, but my specimens from that formation are few, small, and very imperfect; it is abundant in the bed above it. It is in all probability the same as the American species. The ribs extend as far as the junction of the succeeding volution, giving the spire a costated appearance; the striæ or ridges are few, and are carried over the ribs, and the left lip is generally reflected over the umbilicus; and the columella has three folds, the upper one obsolete, sometimes invisible; the lower one forms the edge of the canal; outer lip sharp, and generally free from denticulations, although the ridges are sometimes visible within. This species is constant in its character, and presents very little variation in form; the costæ are generally invisible on the base of the body whorl, and in one or two specimens they are imperceptible upon the spire.

Möller has described the species (Ind. Moll. Grœnl.) under the name *Admete* (*Kröyer*) *crispa;* it differs from the true Cancellariæ, and might probably form a good generic type.

4. CANCELLARIA SUBANGULOSA. *S. Wood.* Tab. VII, fig. 20, *a—b.*

CANCELLARIA SUBANGULOSA. *S. Wood.* Catalogue 1842.

C. Testá minimá, fusiformi, spirá elevatá, apice acuto, longitudinaliter costellatá, transversim striatá, reticulatá; anfractibus quinque, supernè subangulatis; canali brevissimá; columellá triplicatá.

Shell small, fusiform, with an elevated spire, and acute apex, longitudinally costulated, and transversely striated; volutions five, subangular at the upper part; aperture ovate, slightly acuminated at the base ; left lip reflected, with a small open umbilicus, and three very minute folds upon the columella.

Axis, $\frac{3}{16}$ of an inch.

Locality. Cor. Crag, Sutton.

But a few specimens of this species have come into my possession. It appears to be distinct from the preceding one in the angular form of volution, and in being more elongated, with a slight inflection of the outer lip at the angle of the volution. In all my numerous specimens of *C. costellifera* there are none presenting the angular form of volution so distinct in this shell ; it is nevertheless possible it may be only a variety of that species.

Two worn specimens from the Red Crag, in my cabinet, much resemble the *C. læviuscula* of ' Min. Conch. ;' they are in a mutilated condition, and may probably have been introduced from the London clay.

TRICHOTROPIS, *Broderip* and *G. B. Sowerby.* 1829.
TRICHOPHERE. *Desh.* 1830.
TRICHOPODUS. *Swains.* 1840.

Gen. Char. Shell turbinated or subfusiform, with spiral ridges or carinæ upon the volutions, generally thin ; spire more or less elevated, with an acute apex ; aperture subovate, acuminated at the base ; columella obliquely truncate and flattened, with an open umbilicus. In a recent state it is covered with an epidermis, and has a corneous operculum, with a lateral nucleus.

A few species only of this genus are yet known, and those are boreal forms. The name was given from the hairy keels (θριξ, τριχὸς, hair, and τροπις, keel) which surround the species, forming the type of the genus. The animal is said to resemble in many of its characters that of the genus Buccinum, connecting it with Cancellaria ; the shell, however, differs from the latter in having a form of aperture like that of Purpura, with a flattened columella, and also in the absence of oblique folds.

1. TRICHOTROPIS BOREALIS. *Brod.* and *Sow.* Tab. VII, fig. 17, *a—b*, and
Tab. XIX, fig. 11, *a—b*.
TRICHOTROPIS BOREALIS. *Brod.* and *Sow.* Zool. Journal, 1829, vol. iv, p. 375.
— *Gould.* Inv. of Massach. p. 300. 1841.
TRICHOTROPIS ACUMINATA. *Jeffreys.* MS.
— COSTELLATUS. *Couthouy.* Bost. Journ. Nat. Hist. vol. ii, p. 108, pl. 3, fig. 2.
FUSUS UMBILICATUS. *Smith.* Mem. of Wern. Nat. Hist. Soc. vol. viii, p. 50, pl. 1, fig. 2, 1838.

Tr. Testá ovato-fusiformi; spirá elevatá, apice acuminato; anfractibus sex, convexis, cingulatis, supernè planiusculis; longitudinaliter tenuissimè striatis; aperturá ovatá, sub-canaliculatá; labio complanato; umbilico profundo.

Shell ovato-fusiform, with six deeply-defined volutions, and an elevated spire ; whorls rather flattened above, ornamented with four or five elevated transverse, spiral ridges, nearly equidistant; and with fine longitudinal striæ or lines of growth; aperture subcanaliculated; left lip slightly reflected and flattened; umbilicus open and deep.

Axis, ½ an inch.

Locality.   Cor. Crag, Sutton.

Mam. Crag, Bridlington.   Recent, Oban.

I have as yet found but two or three specimens of this species in the Coralline Crag, and these strongly resemble a recent shell in my possession obtained from the Bay of Oban.   In the fossil the spaces between the ridges are quite flat, with fine visible lines of growth; the recent shell has sometimes a small intermediate ridge, but these differences are not constant. Specimens from Bridlington, from the cabinets of Messrs. Bean and Leckenby, are of a less elongated form, and are thicker and stronger, with a very flattened and purpura-like form of inner lip, and having a more open umbilicus, with less prominent transverse ridges; I believe it, however, to be only a variety. Fig. 11, Tab. xix, represents a specimen belonging to Mr. Leckenby which differs from the recent form considerably more than the shell from the older formation.

CERITHIUM,* *Adanson.* 1757.

BITTIUM.   *Leach.* MS. 1819.
POTAMIDES.   *Brongn.*
TRIFORIS.   *Desh.* 1824.
POTAMIDUM.   *Flem.* 1828.

Gen. Char.   Shell elongate, or turriculate, with an elevated or pyramidal spire, composed of numerous volutions; ribbed, striated, or tuberculated, occasionally smooth; aperture subquadrate, terminating in a short recurved canal; outer lip more or less sinuated, sometimes expanded; operculum corneous.

In this genus Lamarck included a large number of species, both marine and fluviatile, the latter of which have been separated and placed in a distinct genus by Brongniart under the name of Potamides.   There is no difference whatever in the shells themselves; a species from the purely fresh-water formations of Hordwell has a deep and recurved canal, which is the character generally considered as peculiar to the marine species.   Many of these species are inhabitants of estuaries, and capable of enduring an existence either in fresh or salt water.   The same species in the Mammaliferous Crag is found with land and fresh-water shells, which in the Red Crag is associated with purely marine forms.

The greater number of species constituting this genus have a distinct and well-marked canal at the base of the aperture, which, however, in some species, dwindles into a slight inflection of the lip at the lower part of the columella, thus approaching in appearance some of the shells in the genus Turritella.

* The derivation of this genus is not well determined.   Hermansen considers it probably from κηρίς, a kind of fish, or from κηρυκίον, *buccinulum.*   Agassiz derives it from κεράτιον, a little horn.

1. **Cerithium tricinctum** (?). *Broc.* Tab. VIII, fig. 1 *a—b,* and 2.

    Murex tricinctus. *Broc.* Conch. foss. Subap. pl. 9, fig. 23, 1816.
    Cerithium tricinctum. *Nyst.* Coq. foss. de Belg. p. 539, pl. 42, fig. 7, 1844.
    — *Dujardin.* Mém. Soc. Géol. de France, tom. xi, p. 288, 1837.
    Cerithium punctatum. *Woodward.* Geol. of Norf. t. 3, fig. 29, 1833.
    — *S. Wood.* Catalogue 1842.

C. Testá elongato-turritá, æqualiter granulata; anfractibus numerosis, planulatis, triplici serie granulorum cinctis, columellá sine plicatá.

Shell elongato-turreted; spire subulate and elevated; volutions flat, ornamented with three equally granulated bands; aperture subquadrate, outer lip curved; columella without a fold.

Axis, 2 inches.

Locality. Red Crag, *passim.*

    Mam. Crag. Thorpe and Bramerton.

Although this species is common enough in the Red Crag, all the specimens I have seen are in bad condition. In the Subapennine shells, and generally in those from Touraine, there is a distinct fold upon the columella above, and independent of that projection which forms one side of the basal canal. It is possible it may be worn away in the Crag shells, as, in some specimens from the Touraine beds, which were given me by Mr. Lyell, this character has quite disappeared, and it was by that alone it could be distinguished. In more perfect specimens, striæ or ridges on the base of the volutions may be observed, which become larger toward the outer edge, like those upon the Touraine shells. Specimens are sometimes found with four bands, while others have only two. Some are so much eroded and altered as to have the volutions distinctly convex.

Fig. 1 (*a*) is from the Mam. Crag, Thorpe.

1 (*b*) from the Red Crag, Sutton.

2 is an eroded specimen, with very convex volutions, and is probably only an altered form of this species.

2. **Cerithium variculosum.** *Nyst.* Tab. VIII, fig. 3, 3 *a.*

    Cerithium variculosum. *Nyst.* Coq. foss. de Belg. p. 540, pl. 42, fig. 9, 1844.
    Cerithium punctulum. *S. Wood.* Catalogue 1842.

C. Testá clongatá, turritá, subulatá; apice acuminatá; anfractibus 9—10, convexiusculis; longitudinaliter plicatis; transversim quadristriatis; striis elevatis, tuberculato nodosis, labro acuto.

Shell elongate, turreted, and regularly tapering, with an acuminated apex; whorls 9—10, slightly convex, covered with four granulated bands; outer lip sharp; base emarginate, without a canal.

Axis, ½ an inch.

Locality. Red Crag, Walton Naze.

Of this species I have but three or four specimens, and those are not quite perfect. It is presumed to be the same as the Belgian species, though the aperture in the figure referred to above is very unlike that of our shell, and is probably not well represented. It appears to differ from *C. lima,* Brug., in not having the occasional obtuse varices of that shell, and the volutions are not so flat as in that species. The base of my specimen is smooth, but perhaps it is not so when perfect. *Cerith. lacteum,* Phil., is described as having but three rows of granules; our shell has four, and they are very distinct.

The name in my Catalogue being without description, though prior to that of M. Nyst, must of course remain only as a synonyme.

3. CERITHIUM TRILINEATUM. *Phil.* Tab. VIII, fig. 4, 4 a.*

CERITHIUM TRILINEATUM. *Phil.* Enum. Moll. Sic. vol. i, p. 195, t. 11, fig. 13, 1836.
—   *Dujard.* Mém. de la Soc. Géol. de France, p. 289, 1837.
—   *S. Wood.* Catalogue 1842.

C. Testá minutá, turritá, subulatá; apice obtuso, costato, et cancellato; anfractibus planiusculis, cingulos tres obtusos ferentibus; interstitiis longitudinaliter striatis; canali brevi.

Shell small, turreted, and tapering, with an obtuse apex; volutions numerous, and rather flat, ornamented, with three transverse obtuse ridges, having longitudinal striæ between them; aperture subquadrate; canal short and inflected.

Axis, ⅜ of an inch.

Locality. Cor. Crag, Sutton.   Recent, Mediterranean.

Imperfect specimens of this species are by no means rare. The shell is thin and fragile. It is peculiarly distinguished by three transverse bands or ridges not quite so broad as the spaces between them: these are obtuse or rather flat upon the top, with longitudinal striæ or visible lines of growth in the interspaces, and there is a smaller ridge at the base of the volution, below which it is smooth. The apex is very obtuse. The first volution is distinctly marked with numerous longitudinal costæ, but is without the transverse ridges, while on the second it is strongly cancellated, being probably the peculiar markings of the young shell previous to its exclusion from the egg.

4. CERITHIUM TUBERCULARE. *Mont.* Tab. VIII, fig. 5, a—c.

Fig. 5, var. VULGARIS. *S. Wood.*
5 a, Magnified portion.
5 b, var. SUBULATUM.
5 c, var. NANUM.
MUREX TUBERCULARIS. *Mont.* Test. Brit. p. 270, 1803.
CERITHIUM PYGMÆUM (?). *Phil.* En. Moll. Sic. t. 25, fig. 26, 1844.
—   HENCKELII. *Nyst.* Coq. foss. de Belg. p. 340, pl. 41, fig. 12, 1844.
—   TUBERCULARE. *S. Wood.* Catalogue 1842.

* Fig. 4 a is a highly magnified representation of the apex.

C. Testá minimá, cylindraceo-turritá ; anfractibus planiusculis ; cingulos tres granosos ferentibus ; basi lævi, canali brevissimo.

Shell small, subcylindrical, and turriculate ; volutions flattish, covered with three transverse granular ridges ; base naked and smooth, with very short canal.

Axis, ¼ of an inch.

Locality. Cor. Crag, Sutton. Recent, British Seas.

An abundant shell, and very variable in its proportional dimensions : some are tapering and elongate, exceeding in length its diameter at least four times ; while others are nearly half as broad as they are long. It is distinguished by having three sharp ridges of tubercles, and a fourth ridge, which is smaller, without the tubercles, at the base, or rather edge, of the last volution. A canal is formed at the suture by the elevation of the tubercular ridges. The base is generally naked, smooth, and flat ; but in some there is a thick and elevated spiral ridge, behind which is a deep sulcus. *Cer. pygmæum,* Phil., looks like one of its varieties, and the Belgian shell I imagine, from the figure, to be the same. I have fragments which indicate a length of at least three eighths of an inch.

5. CERITHIUM METAXA (?). *Delle Chiaje.* Tab. VIII, fig. 6, 6 *a.*
> MUREX METAXA. *Del. Chi.* Mem. 1826, p. 211, t. 49, fig. 29-31.
> CERITHIUM CREPERUM. *S. Wood.* Catalogue 1842.

C. Testá minimá, cylindraceo-turritá ; anfractibus convexis, numerosis; cingulos granulosos quatuor ferentibus ; basi lævi ; labro acuto ; canali brevissimo.

Shell minute, turriculate, and subcylindrical ; whorls numerous, convex, ornamented with four transverse granulated ridges ; base smooth ; outer lip sharp, with very short or emarginate canal.

Axis, ⅜ of an inch.

Locality. Cor. Crag, Sutton.

This is not a very abundant shell, and the specimens are generally broken. It appears, however, to differ specifically from *C. tuberculare* in always having four ridges of granules upon the spire, with a smaller one around the base ; the form of the volution is convex, with the lower portion projecting a little, which gives it a subconoidal form. It appears to correspond with the description given by Della Chiaje, but I am not acquainted with the shell he describes. It differs also, I conceive, from *C. variculosum* in being much more elongate, and by having a deeper suture, and fewer costæ. In *C. variculosum* there are eighteen costæ in the last volution ; in this they are from ten to twelve. In the figures these characters are not sufficiently distinguished.

6. CERITHIUM CRIBRARIUM. *S. Wood.* Tab. VIII, fig. 7, *a—b.*
> CERITHIUM CRIBRARIUM. *S. Wood.* Catalogue 1842.

C. Testá minimá, subulatá, elongatá, tenui ; anfractibus numerosis, convexis, cingulis quatuor ornatis, longitudinaliter decussatis ; basi lævi ; canali brevissimo.

Shell minute, tapering, and elongate, thin, and fragile, with numerous convex volutions, ornamented with four or five elevated transverse ridges, decussated by elevated lines of growth; base naked; canal very short.

Axis, ⅜ of an inch.

Locality.   Cor. Crag, Sutton.

A few imperfect specimens of what I had considered in my Catalogue as a distinct species are in my cabinet; they appear to differ from the preceding in the absence of the granules upon the transverse bands; it is possible it may be only an altered form, although at least twenty fragments present the same characters; but better specimens are necessary for such a determination.   Its present name may therefore be considered only as provisional.

7. CERITHIUM PERPULCHRUM. *S. Wood.* Tab. VIII, fig. 10, 10 *a*.

C. Testá parvá, conico-turritá, vel subulatá; spirá elevatá; apice acuminato; anfractibus 10—11, *convexiusculis, cingulos tres vel quatuor obtusos ferentibus, interstitiis tenuissimè striatis, longitudinaliter obsoletè plicatis; labro subsinuato; canali brevissimo.*

Shell small, elongato-conical, or turriculate, with an elevated spire, and acute apex; volutions slightly convex, furnished with three or four obtuse transverse ridges, having fine striæ between them, obsoletely plicated longitudinally; canal very short; outer lip subsinuated and slightly recurved, with a somewhat expanded inner lip.

Axis, ½ an inch.

Locality.   Cor. Crag, Gedgrave.

One specimen in my own cabinet, and two from that of Mr. Daniel, are all that I have seen, and these are somewhat in an altered condition, with a slightly eroded surface.   It bears a resemblance to *C. turritellatum*, Deshayes (Coq. foss. des Env. de Paris, p. 415, pl. 49, f. 10-11), but is much more tapering in form, with rather less convex volutions, and has more distinct longitudinal plicæ, produced by a slight reflection of the outer lip; the two ridges of the lower part of the volution are the most prominent, particularly the one around the edge of the base.   The lower part of the outer lip is a good deal expanded, and the inner lip considerably spread.   It is a pretty shell, and I was unwilling it should be unfigured, but, like the preceding one, better specimens are necessary for correct determination.

SECT. β.   SINISTRAL.

8. CERITHIUM ADVERSUM. *Mont.* Tab. VIII, fig. 8, 8 *a*.

MUREX ADVERSUS. Mont. Test, Brit. p. 271, 1803.

CERITHIUM ADVERSUM. Brown. Illust. Brit. Conch. pl. 48, fig. 64 (male), 1827.

—   S. Wood. Catalogue 1842.

TRIFORIS ADVERSA. Lovén. Ind. Moll. Scand. p. 21, 1846.

C. Testá minutá, subcylindricá, sinistrorsá; anfractibus planis, cingulos tres granosos ferentibus; apice acuto; canali brevi recurvo, subclauso.

Shell small, subcylindrical, especially the lower half; volutions flat, with three granulated ridges; apex acute; canal short, recurved, and nearly closed.

Axis, $\frac{5}{16}$ of an inch.

Locality. Cor. Crag, Sutton.   Recent, Coast of Britain.

This species is not by any means rare in the Coralline beds, though I have not yet seen it from the Red Crag. There is, I think, no doubt of its identity with the recent British shell. It has commonly three granulated ridges on the body of the whorl, the middle one of which is generally the smallest; and sometimes there are only two, with two plain ones on the base of the last volution. The last whorl is rather contracted, which gives a cylindrical form to the lower half of the shell. A kind of canal is formed at the suture by the elevation of the ridges. The mouth is often broken, and several of my specimens show that it has been repaired by the animal. When perfect, the lower part of the outer lip is a little expanded, giving the mouth a subquadrate form, with a deep sinus in the upper part at the suture. The canals are never quite closed at either extremity of the aperture in any of my specimens.

9. CERITHIUM GRANOSUM. *S. Wood.* Tab. VIII, fig. 9.

    CERITHIUM GRANOSUM. *S. Wood.* Catalogue 1842.

C. Testá subulato-turritá, sinistrorsá; apice obtuso; anfractibus convexiusculis, transversim quadri vel quinque striatis, striis elevatis, tuberculato-granosis; canali brevi, recurvo.

Shell elongate, turriculate, and tapering; whorls sinistral, slightly convex, covered with four or five granular ridges, aperture subcircular; canal short and recurved.

Axis, $\frac{3}{4}$ of an inch.

Locality. Cor. Crag, Sutton.

    Red Crag, Walton Naze.

Imperfect specimens of this shell are by no means rare, but I have nothing that will show the mouth entire. It appears so very different from the figure and description of *C. sinistratum,* Nyst (Coq. foss. de Belg. p. 541, pl. 42, f. 10), that I cannot imagine the two to belong to the same species, unless the Belgian shell has undergone a material alteration by the removal of the granular ridges. The volutions of the Crag shell are rather more conoidal than convex, with four or five nodose ridges when the shell is perfect; the lower one rather projecting, with a wider space above the suture: my best specimen has twelve volutions and an obtuse apex. The first two whorls differ from the others in being sharply costated longitudinally, and finely striated transversely. The base in all my specimens is free from striæ. When the granules are worn down, the surface of the shell appears regularly cancellated.

SECT. HOLOSTOMATA. FLEM.

TURRITELLA,* *Lam.* 1815.
TURBO (spec.) *Linn.*

Gen. Char. Shell elongate and turriculate, generally thick and strong, tapering to an acute apex, volutions numerous; aperture entire, subcircular, or slightly quadrate; outer lip thin and sharp, somewhat sinuous; operculum corneous and spiral.

Shells of this genus are generally well distinguished from those of any other, excepting it be perhaps some aberrant forms of Cerithium; a more or less emarginate base distinguishes the latter genus, although in some species this character is very imperfectly displayed.

As a genus this is not restricted to any climate, some species being found within the tropics, while others are inhabitants of the coasts of Greenland and Scandinavia, and, according to Mr. J. Sowerby, it appears in the Protozoic rocks (vide Sil. Syst.); these shells, however, resemble so much in form those of *Murchisonia*, that unless the specimens are very perfect it would be difficult to distinguish them. There is, however, a well-marked species from the Greensand formation. (Vide Min. Con. t. 565.)

1. TURRITELLA COMMUNIS. *Risso.* Tab. IX, fig. 9, *a —b.*
   TURBO TEREBRA. *Mont.* Test. Brit. p. 293, 1803.
   TURRITELLA COMMUNIS. *Risso.* Hist. Nat. des princ. Prod. de l'Europe, iv, p. 106, fig. 37, 1826.
   — TEREBRA. *J. Sow.* Min. Conch. t. 565, fig. 3, 1827.
   — — *Woodward.* Geol. of Norf. p. 44, 1833.
   — LINNÆI. *Desh.* Exp. de Morée, tom. iii, p. 146.
   — — *Dujard.* Mém. de la Soc. Géol. de France, p. 287, 1837.
   — TEREBRA. *S. Wood.* Catalogue 1842.
   — COMMUNIS. *Phil.* En. Moll. Sic. vol. ii, p. 160, 1844.
   — PLEBEIA. *Say.* Lyell in Proc. of Geol. Soc. 1845, vol. iv, pt. 3, p. 554.

T. Testá turritá, subulatá; anfractibus convexiusculis; transversim lineatis; lineis 8—10, elevatis acutiusculis; æqualibus, vel inæqualibus, cinctis.

Shell turriculate and tapering, with very slightly convex volutions; covered with 8—10 fine ridges, sometimes equally distributed, at others rather irregular in size and number; outer lip sinuous.

Axis, 1¾ inch.

Locality. Red Crag, Sutton.

Mam. Crag, Bramerton and Bridlington.   Recent, British Seas.

This shell is given by Professor E. Forbes, in his 'Report upon the Existing Fauna and Flora of the British Isles,' as a Coralline Crag species. I have no specimen from that formation. It is, however, very probable it may have lived in the seas which deposited the Coralline Crag, having been given as a Touraine species by M. Dujardin, and as an

* Etym. From *Turris,* a tower.

American Miocene fossil by Mr. Lyell. The striæ or ridges are rather irregular, though generally there is one large and one small one alternately ; they are all sharp and angular. The ridges upon recent specimens are also irregular in their disposition, having sometimes fine transverse striæ between them. Fig. 9 *b* is from the Red Crag, resembling the general form of the recent shell. Fig 9 *a* is from the Mammaliferous Crag at Thorpe, which I presume to be the same species, although it is more subulate in form, with a deeper suture, and more convex volutions.

2. TURRITELLA IMBRICATARIA. *Lam.* Tab. IX, fig. 10, *a—b.*

TURRITELLA IMBRICATARIA, *Lam.* An. du Mus. vol. iv, p. 216, pl. 37, fig. 7, 1804.
—   *Desh.* Coq. foss. des Env. de Par. pl. 35, fig. 1-2, and pl. 37, fig. 9-10.
—   *Nyst.* Coq. foss. de Belg. p. 397, pl. 37, fig. 5.
TURRITELLA CONOIDEA. *J. Sow.* Min. Conch. t. 51, fig. 145, 1814.
—   *S. Wood.* Catalogue 1842.

T. Testá, elongato-turritá, subulatá; anfractibus planatis, ad basim subangulatis, transversim striatis, striis imbricatis, vel subtilissime granulatis.

Shell elongate, turreted, and tapering, with flattened volutions, slightly angulated at the lower part ; transversely striated with imbricated or finely-granulated striæ.

Axis, 2 inches.

Locality. Red Crag, Sutton and Newbourn.

My cabinet contains about thirty specimens of this shell, which is no doubt the same species as the one so abundant in the Eocene formation, at Barton ; my specimens are all more or less rubbed, and may probably be a diluvial introduction from the older bed, although that number is rather large for such a casualty. One specimen is covered with fine and regular transverse striæ, which though rubbed show the remains of granulations ; the angular form of the base of the volution appears to be its distinguishing character. Fig. 10 *b* is from a specimen much worn and eroded, and exhibits a more conical form of volution than the generality of these shells.

3. TURRITELLA INCRASSATA. *J Sow.* Tab. IX, fig. 7, *a—d.*

TURRITELLA INCRASSATA. *J. Sow.* Min. Conch. t. 51, fig. 6, 1814.
TURBO TRIPLICATUS. *Brocchi.* Foss. Subapenn. t. 6, fig. 14, 1814.
— DUPLICATUS (?). - - - -   t. 6, fig. 18.
— IMBRICATARIUS. - - - -   t. 6, fig. 12.
— VERMICULARIS. - - - -   t. 6, fig. 13.
TURRITELLA DUPLICATA. *Dubois de Montp.* Conch. foss. Wolhyn. Podol. pl. 2, fig. 19-20, 1831.
— TRIPLICATA. *Dujard.* Mém. Soc. Géol. de France, tom. ii, pt. 2, p. 287, 1837.
— — *Grateloup.* Foss. de Dax (Actes de la Soc. Linn. de Bord.), tom. v, p. 161, 1838.
— — *Nyst.* Coq. foss. de Belg. pl. 37, fig. 7-8, 1844.
— INCRASSATA. *S. Wood.* Catalogue 1842.
— BICINCTA. - - - -   var. β, fig. 7 *d.*

T. Testá turritá, anfractibus planulatis; transversim striatis, cingulos duos vel tres obtusos distantes ferentibus; cingulo intermedio crassiore, supremo obsoleto.

Shell turriculate, with flat volutions, subcarinate at the base; finely striated transversely, and banded at rather irregular distances, with two or three obtuse ridges, the upper one nearly obsolete.

Axis, 2½ inches.

Locality.   Cor. Crag, Gedgrave and Ramsholt.

                Red Crag, *passim.*   Recent, Mediterranean.

This species is abundant, and very variable in regard to the number and arrangement of its banded ridges. That which I considered as a distinct species, and named *bicinctus* (Tab. IX, fig. 7 *d*) when my Catalogue was published, I have now reason to believe is only a variety. The young or upper volutions of most specimens have the banded ridges more prominent and distinct than upon the larger part of the shell. Fig. 7 *a* appears to correspond with Brocchi's figures (Tab. VI, figs. 12 and 13), which are, I imagine, only varieties with more elevated ridges.

Figs. 7 *a*, *d*, are from the Coralline Crag.

Figs. 7 *b*, *c*, are from the Red Crag.

4. TURRITELLA PLANISPIRA. *S. Wood.* Tab. IX, fig. 11.

       TURRITELLA PLANISPIRA.   *S. Wood.* Catalogue 1842.
            —   *Nyst.* Coq. foss. de Belg. p. 401, pl. 38, fig. 9, 1844.

T. Testá turritá, subulatá; anfractibus planatis; transversim striatis; striis regularibus sub-æquidistantibus.

Shell turreted and tapering, with nearly flat volutions; transversely striated; striæ regular, nearly equidistant.

Axis, 1½ inch.

Locality.   Cor. Crag, Sutton.

I have only been able to procure fragments of this species, and they appear to agree with the shell figured by M. Nyst. The whorls are very flat, with scarcely a distinct suture in some specimens. In the young state the middle ridge is always the most prominent, taking off from the flatness of the volution, which distinguishes it when further increased. The ridges or striæ are generally regular, though sometimes alternating, one large and one small. M. Nyst, as well as myself, have considered this distinct, and it is rather singular that we should both have chosen the same name for it. It is, however, possible this may be only a variety of *T. communis*, but I have never seen that shell with such perfectly flat volutions, or so subulate in form.

5. TURRITELLA CLATHRATULA. *S. Wood*, 1847. Tab. IX, fig. 6, 8 *a*.

C. Testá turritá, subulatá, anfractibus novem convexiusculis; lineis quinque elevatis tranvsersis cinctis; interstitiis striatis vel clathratis; aperturá subquadratá.

Shell turreted and tapering, with about nine slightly convex volutions, ornamented with five transverse rather rounded ridges; the sulci between them longitudinally striated or imbricated; aperture subquadrate.

Axis, ⅜ of an inch.

Locality. Mam. Crag, Bridlington.

The specimen figured is from the Museum of the Philosophic Society of York, and was obtained at Bridlington, where, I am informed by Mr. Charlesworth, who sent the specimen, it is not very rare. It much resembles a recent and undescribed species, said by Mr. G. B. Sowerby, sen., to be from the north-west coast of Africa.

Its distinguishing character is the number of ridges upon the volution. These are placed at regular distances, the lower one projecting a little, which gives the whorl a slight conoidal form ; the lines of growth are rough and imbricated, and most distinct between the ridges, which are somewhat rubbed and rounded. A subquadrangular form of volution gives it an angular edge and a flat base, which has three or four spiral striæ or ridges, the outer ones most prominent, with elevated lines of growth. The columella is nearly straight, with a subcanaliculated base, or everted left lip, bringing this species into close approximation with Cerithium. It is one of those shells by which the two genera might be connected.

A fragment of a specimen of Turritella from the same locality, apparently a distinct species, was given to me by my friend F. Edwards, Esq., though it is too imperfect for figuring or description.

PYRAMIDELLA,* *Lam.* 1815.

Gen. Char. Shell turreted, elongato-pyramidal, generally smooth and glossy, sometimes costated, composed of numerous closely-set volutions ; aperture subcanaliculated ; columella nearly straight, with several folds or ridges.

1. PYRAMIDELLA LÆVIUSCULA. *S. Wood.* Tab. IX, fig. 2, *a—b.*

       PYRAMIDELLA UNISULCATA (?). *Dujardin.* Mém. Soc. Géol. de France, t. 11, 1837.

       PYRAMIDELLA LÆVIUSCULA. *Grateloup.* Bord. foss. pl. 7, fig. 79, 80, 1838.

       —   TEREBELLATA. *S. Wood.* Catalogue 1842.

       —   —   *Nyst.* Coq. foss. de Belg. p. 431, pl. 37, fig. 28, 1844.

P. Testá elongato-turritá, lævigatá, politá, apice obtuso; anfractibus 8—9 *planatis; suturá subsulcatá; aperturá subovatá; labro acuto, intus dentato; columellá rectá; triplicatá, plicis inæqualibus.*

Shell elongato-turriculate, smooth, and glossy, with an obtuse apex ; whorls flat, about nine in number ; suture distinct, deep, subcanaliculated ; aperture subovate ; outer lip sharp, toothed within ; columella straight, slightly replicate, with three unequal folds.

Axis, 5/16 of an inch.

Locality. Cor. Crag, Sutton.

A very abundant species. This shell appears to be identical with the Belgian

* Etym. *Pyramis,* pyramid or spire.

fossil, but I think distinct from the Paris basin shell, *P. terebellata*, Desh. (Coq. foss. des Env. de Paris, pl. 22, f. 7–8.)   In the Crag shell, the upper fold upon the columella is large, sharp, and elevated, with very little obliquity ; below it are two small oblique folds, nearly equal sized, the lower one forming the base of the columella, and producing there an incipient sinus ; the outer lip is sharp, and it has about half a dozen obtuse teeth within it, which are not continuous, but produced at intervals.

It is nevertheless possible that this may be the same as the Eocene species, but my specimens are all so much less, as well as differing in other characters, that I have left it with my provisional name until more accurate comparisons can be made with the foreign shell.

CHEMNITZIA,* *D'Orb.* 1839.

TURBO (spec.) *Mont.*

TURBONILLA. *Leach,* 1819.

PARTHENIA. *Lowe,* 1842.

ORTHOSTELIS. *Arad.* and *Magg.* (ex *Phil.*)

PYRGISCUS. *Philippi.*

EULIMA (spec.) *Philippi.*

LOXONEMA (?). *Phillips.*

MELANIA. *Scacchi.*

EULIMELLA. *Forbes.*

Gen. Char. Shell turriculate, with numerous volutions, nearly flat, having a distinct suture, aperture subquadrate, columella straight, with or without a fold ; apex mammillated by the reversed position of the extreme spire, exterior generally striated or costated, sometimes smooth and plain.

Dr. Leach, in MS., united a group of shells with these characters, to which he gave the name of Turbonilla, which was adopted by Risso in 1826.   In June, 1840, the Rev. T. Lowe read a paper before the Zoological Society, describing several of these species, and proposed for them the generic name of Parthenia, considering that given by Dr. Leach as being incorrect in composition, and otherwise preoccupied by another genus of Mollusca.   In 1840, according to M. Philippi, Wiegman proposed the name of Pyrgiscus for some marine Melaniæ, which it is probable was intended for these shells.   In 1839, M. d'Orbigny published some species under the above name (in Webb and Berth. Hist. Nat. des Iles Canar. Moll.), and this name appears to have the precedence of all but that of Leach.   M. Lovén, in his Synopsis, has restored the name of Turbonilla, which perhaps, after all, notwithstanding the objection by Mr. Lowe, ought to be the one employed.

In the young state, probably before its exclusion from the egg, the first volution appears to have taken a sinistral form, and afterwards to have changed into a dextral direction, and the apex is rendered obtuse by the folding down or horizontal position of its primary volution.

* Etym. After Chemnitz.

Shells somewhat of this form have been long known from the older Secondary Formations, and were placed in a genus by Phillips, under the name of Loxonema; they do not, however, appear to possess the quadrate form of aperture and straight columella of these shells, although they have been united to this genus by M. d'Orbigny and other continental authors.

As a recent genus it appears to have a very extended geographical range, and is rather inclined to deep water.

1. CHEMNITZIA CURVICOSTATA. *S. Wood.* Tab. X, fig. 1, 1 *a.*
TURBONILLA CURVICOSTATA. *S. Wood.* Catalogue 1842.

Ch. Testâ elongato-turritâ, tenui, fragili; anfractibus planiusculis, plicatis; plicis confertis, obliquis, vel curvatis; suturis profundis, labro acuto.

Shell turreted and elongate, thin and fragile, with an obtuse apex; volutions 7—8, rather flat; suture distinct and deep; longitudinally costated, or plicated; costæ oblique and curved; columella straight and naked.

Axis, ⅟₇ of an inch.

Locality. Cor. Crag, Sutton.

I have but three specimens of this species, and those are not in very good condition. The remains of its curved costæ and straight columella are almost the only distinguishing characters. It may possibly be the *Turbo indistinctus,* Mont. (Sup. p. 129), but that shell is described as punctured in the furrows, which I cannot observe in my specimens, the costæ are much curved, broad, and close, and have the spaces between them thin and narrow.

It somewhat resembles the figure of *Ch. terebellum,* Phil. (En. Moll. Sic. vol. ii, p. 138, t. 24, f. 12), but our shell seems to have been larger, and the costæ more oblique and curved. I have therefore, for the present, left it with its provisional name until it can be better determined. The apex in my specimens is very obtuse, flattened, and depressed, but not reversed like the general character of this genus.

2. CHEMNITZIA RUFA. *Phil.* Tab. X, fig. 2, 2 *a,* 2 *b.*
MELANIA RUFA. *Phil.* En. Moll. Sic. vol. i, p. 156, t. 9, fig. 7, 1836.
PARTHENIA CRENATA. *Lowe.* Proc. Zool. Soc. p. 41, 1840.
TURBO CRENATUS. *Montague* (inedit.)
TURBONILLA RUFA. *S. Wood.* Catalogue 1842.
— — *Lovén.* Ind. Moll. Scand. p. 18, 1846.
CHEMNITZIA RUFA. *Phil.* En. Moll. Sic. vol. ii, p. 136, 1844.

Ch. Testâ turriculâ; anfractibus planiusculis; longitudinaliter costatis; costis rectis; interstitiis transversim striatis; suturis profundis; aperturâ subquadratâ; columellâ rectâ.

Shell turriculate; whorls 8—9, flattish; longitudinally costated; ribs vertical, transversely striated; suture deep; aperture subquadrate; outer lip simple, plain within.

Axis, ¼ of an inch; *diameter,* ¼ of its length.

Locality. Cor. Crag, Sutton. Recent, Mediterranean.

My cabinet contains about half a dozen specimens of this species, which appear to correspond with the figure and descriptions above referred to. The volutions are rather flat, but project a little, which leave a slight shoulder at the upper part, and cause the sutures to be deep and distinct. The transverse impressed striæ, or rather sulci, are about six or seven in number on each whorl, and these are continued over the base. The costæ are nearly vertical and straight, and from eighteen to twenty in number on the last volution. The enlarged drawing in the very able work by M. Philippi, referred to above, is a good representation of our shell. The costæ in our figure are scarcely erect enough.

3. CHEMNITZIA COSTARIA. *S. Wood.* Tab. IX, fig. 3, 3 *a.*
TURBONILLA COSTARIA. *S. Wood.* Catalogue 1842.

Ch. Testá turritá, subulatá; anfractibus numerosis, convexiusculis; longitudinaliter costatis; costis obliquis, obtusis, confertis; transversim sulcatis; labro incrassato, intus denticulato.

Shell elongate, turriculate, tapering; volutions 8—10; longitudinally costated; costæ oblique, obtuse; transversely striated or sulcated; suture distinct; aperture subquadrate; outer lip thickened, and dentated within.

Axis, $\frac{3}{8}$ of an inch.

Locality. Cor. Crag, Sutton.

This species is by no means rare. It resembles much *Ch. elegantissima,* but is less cylindrical, and is transversely striated or sulcated; the costæ are oblique, inclining a little to the left. It is beautifully grooved or sculptured, the lines being carried over the ribs. The lower part of the outer lip projects a little, so as to give an obliquity to the ribs, which are about twelve in number on the last volution. These ribs terminate at the edge of the base, but the transverse striæ are continued over it. There is no appearance of any fold upon the columella. The inside of the outer lip is dentated. It appears to differ from *Ch. obliquata,* Phil. (vol. ii, p. 137, t. 24, f. 10), in being much larger, and striated. It differs also from *Ch. rufa,* in having fewer costæ, and in those being more inclined; on the last volution they are about thirteen in number.

4. CHEMNITZIA NITIDISSIMA (?). *Mont.* Tab. IX, fig. 4, 4 *a.*
TURBO NITIDISSIMUS. *Mont.* Test. Brit. p. 299, t. 12, fig. 1, 1803.
TURRITELLA NITIDISSIMA. *Flem.* Brit. An. p. 304, 1828.

Ch. Testá pusillá, turritá, subulatá, lævigatá, politá; anfractibus convexiusculis, vel subconoideis, apice obtuso; aperturá subovatá; labro acuto, intus lævigato.

Shell turriculate, tapering, smooth, and glossy; volutions slightly convex, or rather conoidal; apex obtuse; aperture subovate; outer lip sharp, with a deep and distinct suture.

Axis, $\frac{1}{8}$ of an inch.

Locality. Cor. Crag, Sutton.   Recent, British Seas.

My cabinet contains but four specimens of this species, and those are not in very good condition; the identification, therefore, must be considered doubtful. The volutions project a little at the lower part, which give them a conoidal form: Montague says, "terminating in a fine point:" my specimens have the apex very obtuse, so have all the species of this genus.

5. CHEMNITZIA ELEGANTISSIMA. *Mont.* Tab. X, fig. 5, 5 *a*.

    TURBO ELEGANTISSIMUS. *Mont.* Test. Brit. p. 298, t. 10, fig. 2, 1803.
    MELANIA CAMPANELLA. *Phil.* En. Moll. Sic. vol. i, p. 156, t. ix, fig. 5, 1836.
    PARTHENIA ELEGANTISSIMA. *Lowe.* Zool. Proc. 1840, p. 41.
    TURRITELLA ELEGANTISSIMA. *Flem.* Brit. An. p. 303, 1828.
    EULIMA ELEGANTISSIMA. *Risso.* Hist. Nat. de l'Europ. Merid.

Ch. Testá elongatá, turritá, subulatá, nitidissimá; anfractibus numerosis, convexiusculis, longitudinaliter costatis; costis obliquis, obtusis, confertissimis, interstitiis lævibus; suturis distinctis; labio columellari verticali.

Shell elongate, turreted, tapering, smooth and glossy; whorls very slightly convex, longitudinally costated; costæ numerous, oblique, and obtuse, with a distinct suture: base of volution smooth; aperture subquadrate.

Axis, ¼ of an inch.

Locality. Cor. Crag, Sutton.   Recent, British Seas.

But one specimen of this shell has come into my possession; it appears to be identical with the recent British species. The costæ are not quite vertical, but have a little inclination, so that the lower part inclines to the left of a perpendicular line, but they do not curve. Montague states this species to be covered, in a recent state, with a brown epidermis. It is said by Dr. Fleming (British Animals, p. 303) to be a deep-water shell.

6. CHEMNITZIA INTERNODULA. *S. Wood*, 1847. Tab. X, fig. 6, 6 *a*.

Ch. Testá, subulatá, subcylindricá, elongatá; anfractibus numerosis planis, longitudinaliter costatis; costis obliquis, obtusis, confertis, ad interstitias serie unicá granulorum; aperturá subquadratá; columellá rectá uniplicatá; labro simplici, intus lævigato.

Shell tapering or subcylindrical, smooth, and glossy, with twelve or fourteen flat volutions, longitudinally costated; costæ slightly oblique; interspaces with an elevated nodule; aperture subquadrate; columella straight, with one fold; outer lip simple, smooth within.

Axis, 9/16 of an inch.

Locality. Cor. Crag, Sutton.

           Red Crag, Sutton.

This species is not rare. In some characters it resembles *Turbo gracilis*, Broc. (Tab. VI, fig. 6), but that shell has the ribs more vertical, and has not the connecting link between them. Our shell appears as if its ribs were bound together by something passing through them, which gives a sort of knob between each rib, with a depression

11

above and below it. My specimens are smooth and glossy, but they appear as if they had been rubbed, and the greater number of them are rather more slender than the one represented. The costæ terminate at the edge of the volution, and the base is smooth.

7. CHEMNITZIA FILOSA. *S. Wood.* Tab. X, fig. 7, 7 *a*.

TURBONILLA FILOSA. *S. Wood.* Catalogue 1842.

Ch. Testá turritá, subulatá; anfractibus septem vel octo convexiusculis, longitudinaliter obsoleté costatis; transversim regulariter striatis; basi rotundato; aperturá subovatá; labro extus incrassato, intus lævigato.

Shell turreted and tapering, with seven or eight slightly convex volutions, longitudinally costated; costæ nearly obsolete, transversely striated; aperture subovate; outer lip thickened, varicose, smooth within.

Axis, ½ of an inch.

Locality. Cor. Crag, Sutton. ? Recent, Mediterranean.

I have found about a dozen specimens of this shell, which appear distinct from any species I am acquainted with. It has about eight thread-like striæ equally distributed upon each whorl, which are continued over the obsolete ribs as well as over the base of the shell; the costæ terminate at the edge of the volution, which is somewhat rounded. The outer lip is thickened externally in its adult state, and sometimes the animal has increased the shell beyond its ordinary size, leaving a thickened and obtuse varix upon the exterior. The last volution has from eight to ten erect, obtuse, or varicose costæ. The columella is not so straight in this as in most of the species of the genus, which gives the aperture an ovate form, broader at the lower part.

Some time since Prof. E. Forbes identified this shell with *Parthenia varicosa*, described in his Report upon the Invertebrata of the Ægean Sea, published in the Report of the British Association, 1843. I should have given that name upon his authority, but, upon referring to the description contained therein, I think there must be some mistake; that shell is described as having 11 volutions with 18—20 costæ, and as being only half an inch in length. I have, therefore, for the present, left the species with the provisional name of my Catalogue.

8. CHEMNITZIA DENSECOSTATA. *Phil.* Tab. X, fig. 8, 8 *a*.

CHEMNITZIA DENSECOSTATA. *Phil.* En. Moll. Sic. vol. ii, p. 137, t. 24, fig. 9, 1844.
TURBONILLA CYLINDRELLA. *S. Wood.* Catalogue 1842.

Ch. Testá aciculatá, subcylindraceá; anfractibus convexiusculis, costatis; costis numerosis, elevatis, interstitiis striatis; columellá rectá subplicatá; labro simplici, intus lævigato.

Shell suberect, subcylindrical, slender; whorls slightly convex, costated; costæ numerous, nearly erect, elevated, smooth; intermediate spaces transversely striated; aperture subovate; columella straight, subplicated; outer lip simple, not dentated within.

Axis, ¼ of an inch.

Locality. Cor. Crag, Sutton. Recent, Mediterranean.

This is not a very rare species. I have presumed this to be the shell described by M. Philippi, although my specimens have an obtuse apex, not corresponding in that character with what he calls "apice acuto." The furrows between the ribs are rather wider than the ribs themselves, and run into the suture at the lower part, but not at the upper, from the overlapping of the succeeding volution, and the whorls are rather more convex than those represented at the above reference. The first five or six volutions present an elongato-conical form, after which the shell becomes nearly cylindrical. It is slender and elegant, rather more so than our figure represents, and covered with numerous nearly vertical ribs, the last whorl having as many as twenty-four. I am not acquainted with the recent shell, and have trusted for this identification to the figure and description by M. Philippi.

9. CHEMNITZIA UNICA (?). *Mont.* Tab. X, fig. 9, 9 *a*.
      TURBO UNICUS. *Mont.* Test. Brit. p. 299, t. 12, fig. 2, 1803.
          —   *Turt.* Conch. Dict. p. 209, 1819.
      PARTHENIA PALLIDA (?). *Lowe.* Zool. Proc. p. 42, 1840.
      CHEMNITZIA PALLIDA (?). *Phil.* En. Moll. Sic. vol. ii, p. 136, t. 9, f. 8.
      TURBONILLA SUBULATA. *S. Wood.* Catalogue 1842.

Ch. Testá turritá, subulatá, aciculatá, gracili, politá; anfractibus numerosis, convexius-culis; longitudinaliter plicatis; plicis confertis, obliquis; suturis distinctis; aperturá sub-quadratá; columella rectá.

Shell turriculate, slender, elongate and tapering, smooth and glossy; whorls slightly convex, longitudinally costated; costæ numerous, close, and oblique, intermediate spaces transversely striated (?); suture deep and distinct, with a subquadrate aperture and straight columella.

Axis, ¼ of an inch.

Locality. Cor. Crag, Sutton.   Recent, British Seas.

My cabinet contains five specimens of this slender shell, which is assigned to *Ch. unica* with doubt. It resembles a very slender variety of *Ch. elegantissima*, but differs from that species in being more elongate. The volutions in my specimens project more at the lower part than they do above; and the furrows between the costæ appear to terminate before reaching the suture; whereas, in *elegantissima*, they run into it. My specimens are evidently much rubbed, and do not show the intermediate striæ distinctly. Our shell differs materially from the figure referred to, (*Chemn. pallida*, Phil. t. 9, f. 8,) in being more slender, the costæ more inclined, and the volutions less convex; it has a greater resemblance to *Ch. gracilis* of that author (pl. 24, f. 11), but the furrows in our shell appear to terminate just before reaching the suture, and the costæ incline a little, but do not curve. The aperture is subquadrate, with a straight columella, which has a slight thickening in the middle like an incipient fold.

10. CHEMNITZIA VARICULA. *S. Wood.* Tab. X, fig. 10, 10 *a*.
TURBONILLA VARICULA. *S. Wood.* Catalogue 1842.

Ch. Testá turritá, elongato-pyramidali; anfractibus convexiusculis; longitudaliter costatis; costis 10—12 obtusis depressis, distantibus; transversim striatis; striis vel sulcis 8—9; aperturá subquadratá; labro acuto, intus lævigato.

Shell turreted, elongato-conoidal; whorls slightly convex; suture distinct, longitudinally costated, with small, vertical, and distant costæ, transversely striated or sulcated, with 8—9 depressed lines; mouth subquadrate; outer lip sharp, and smooth within (?).

Axis, ⅜ of an inch.

Locality. Cor. Crag, Sutton.

I have about a dozen specimens, which offer characters resembling this genus, but the mouths of all of them are more or less broken. It slightly resembles *Rissoa varicosa*, Bast., but that is a larger shell, with fewer varices, and is toothed within the outer lip; our shell has 9—10 volutions, is very much less in size, and the costæ become obsolete on the lower portion.

11. CHEMNITZIA SIMILIS. *Forbes.* Tab. X, fig. 11, *a—c*.
TURBONILLA ELEGANTIOR. *S. Wood.* Catalogue 1842.

Ch. Testá turritá, subulatá, elongatá, politá; anfractibus planis; longitudinaliter obsoletè costatis; transversim sulcatis; sulcis quinque; aperturá subquadratá; columellá rectá, obsoletè uniplicato; labro intus lævigato.

Shell elongato-turriculate, tapering, glossy; apex obtuse; longitudinally costated; costæ obsolete; transversely striated or sulcated; suture deep; aperture subquadrate; left lip slightly reflected, with an incipient fold upon the columella; outer lip simple, not dentated within.

Axis, ⅜ of an inch.

Locality. Cor. Crag, Sutton.   Recent, British Seas.

Imperfect specimens of this shell are by no means rare, but the surface is generally much rubbed. It is, however, distinctly marked with about five deep sulci, which cut the numerous and rather obsolete costæ, making the surface somewhat reticulated. The sulci or striæ are continued over the base, which is rather rounded; some have a tooth-like projection upon the columella, in others it is not visible. Professor E. Forbes tells me it is the same as the recent British species, and I have given it upon his authority, Fig. 11, *b*, is the representation of a specimen that has become quite smooth: another specimen in my cabinet has tranverse striæ, but no costæ.*

* Two specimens apparently of another species, belonging probably to this genus, are in my Cabinet, but the mouths are too much broken for correct description.

The minute, sinistral, planorbis-like looking shell figured with my Catalogue (An. and Mag. Nat. Hist. 1842, pl. 5, fig. 12,) is, perhaps, the fry of one of the species of this genus.

ODOSTOMIA,* *Flem.* 1817.

TURBO (spec). *Mont.*
OVATELLA. *Bivona,* 1832.
MELANIA. *Dubois,* 1831.
TURBONILLA. *Lovén,* 1846.

Gen. Char. Shell small, conical, or subulate; generally smooth and glossy, with an obtuse and often a reversed apex; sometimes striated, and occasionally costated longitudinally; aperture ovate, acuminated at the upper part, with a tooth-like projection or fold upon the columella.

This name, as proposed by Dr. Fleming, in the 'Edinburgh Encyclopædia,' art. Conchology, first published in 1817, appears to have been intended for the reception of a number of land shells, with a denticulated aperture, such as Pupa, Clausilia, &c., but was subsequently restricted, or rather transferred by that author in his 'Hist. of Brit. An., 1828,' to a series of small marine shells with an ovate aperture, and a single plait upon the columella.

Mr. Alder has informed me that the animals of this genus have ear-shaped tentacles, with eyes at the internal base, resembling those of *Chemintzia*, to which genus they have been united by M. Lovén. The shells, however, appear to differ from those of the latter genus in being less turriculate, with fewer volutions, and having a less quadrate form of aperture, sufficiently so, I imagine, to be considered as generically distinct. Should it be necessary to unite the two, the name of *Chemnitzia* must yield to the prior one of Dr. Fleming.

1. ODOSTOMIA PLICATA. *Mont.* Tab. IX, fig. 3, *a—b.*

TURBO PLICATUS. *Mont.* Test. Brit. p. 325, pl. 21, 1803.
— CONOIDEUS. *Broc.* Conch. foss. Subapenn. p. 660, c. 16, fig. 2, 1814.
OVATELLA POLITA. *Bivona.* p. 4, t. 1, fig. 7, 1832.
AURICULA CONOIDEA. *Phil.* En. Moll. Sic. vol. i, p. 143, 1836.
TORNATELLA CONOIDEA. *Nyst.* Coq. foss. de Belg. p. 428, pl. 37, fig. 27, 1844.
ODOSTOMIA PLICATA. *S. Wood.* Catalogue 1842.
— — *Thorpe.* Mar. Conch. p. 172, fig. 13, 1844.
— UNIDENTATA (?). - - - p. xxxv, fig. 11.
TURBONILLA PLICATA. *Lovén.* En. Moll. Scand. p. 19, 1846.

Od. Testá crassá, elongato-conoideá, turritá, lævissimá, politá; apice obtuso; anfractibus 7—8 planiusculis; suturis distinctis; ultimo anfractu subangulato; columellá uniplicatá; plicá mediá dentiforme; labro acuto, intus dentato.

Shell thick and strong, elongato-conical, turriculate, smooth and glossy, with an obtuse reversed (?) apex; whorls 7—8, flattish; suture distinct; last whorl subangulated, columella with a sharp and prominent tooth centrally placed; outer lip sharp and dentated within.

* Etym. Ὀδους, a tooth, and στόμα, a mouth.

Axis, $\frac{5}{16}$ of an inch.

Locality. Cor. Crag, Sutton.

    Ramsholt and Gedgrave.   Recent, British Seas.

A very abundant species at Sutton, in the Coralline beds, but I have not met with it from the Red Crag Formation.

My fossil specimens are much larger than any recent individuals that I have seen; but Mr. Alder and M. Lovén consider the fossil as identical with the recent British species, and I have given it upon their authority. It is very variable; some specimens have the volutions subangulated at the base, with a small indistinct ridge in one variety, which has also the umbilicus covered, while, in others, the base is without the angular appearance, and the umbilicus open. The angulated edge of the volution gives a subcanaliculated form of suture to another variety (fig. 3 *a.*) All my specimens are smooth and glossy, with lines of growth occasionally visible.

2. ODOSTOMIA PUPA. *Dubois.* Tab. IX, fig. 5, *a—b.*

   MELANIA PUPA. *Dubois.* Geol. Wolhyn. Podolien, pl. 3, fig. 34-35, 1831.
   ODOSTOMIA PUPA. *S. Wood.* Catalogue 1842.

 Od. Testá minutá, turritá, elongato-conoideá, crassá; anfractibus 6—7 planatis; longitudinaliter plicatis; plicis 12—14 subrectis, confertis; supernè lævigatis, infernè striis duobus, transversis instructis; aperturá ovali; columellá obsoletè, unidentatá.

Shell small and strong, turreted, elongato-conical; whorls flat, longitudinally costated or plicated; folds obtuse, nearly erect, and numerous; with two transverse striæ at the lower part of the volution: aperture ovate, acuminated at the upper part, with an obsolete fold upon the columella.

Axis, $\frac{1}{10}$ of an inch.

Locality. Cor. Crag, Sutton.

Abundant. There is a slight difference between our shell and the figure given by M. Dubois; but, as far as the description goes it appears to correspond; and I presume the two shells to be identical. In his figure the transverse lines are smaller and closer. The Crag shell has a tooth a little above the middle of the columella, very distinct in broken specimens. The outer lip curves at the lower part, giving a slight flexure to the costæ, which are smooth and obtuse, and nearly as wide as the spaces between them. The two transverse lines are visible upon the last volution only, one being hidden in the suture of the spire; these lines are large and rounded, but do not cross the costæ, only bind them, as it were, together, and do not alter the form of the outer lip.

3. ODOSTOMIA PELLUCIDA. *Adams.* Tab. IX, fig. 4, *a—b.*

   TURBO PELLUCIDUS. *Adams.* Linn. Trans. iii, p. 66, t. 13, fig. 33-34.
   — DECUSSATUS. *Mont.* Test. Brit. p. 322, t. 12, fig. 4, 1803.
   — — *Thorpe.* Mar. Conch. p. 169, 1844.
   ODOSTOMIA RETICULATA. *S. Wood.* Catalogue 1842.

 Od. Testá pusillá, elongato-ovatá, pellucidá (?) *tenui, et fragili; apice obtuso;*

anfractibus quatuor vel quinque convexis; suturis profundis; longitudinaliter tenuissimè costellatis; transversim striatis, decussatis; labro acuto; umbilico mediocri; columellâ obsoleté unidentâ.

Shell small, thin, fragile, and pellucid (?) with four or five convex volutions, and a deep and distinct suture; longitudinally costulated with 18—22 slightly raised ribs decussated by seven or eight transverse lines; aperture ovate; outer lip sharp and plain within; a small but distinct umbilicus; and a rather obscure fold upon the columella.

Axis, $\frac{1}{8}$ of an inch.

Locality. Cor. Crag. Sutton.   Recent, British Seas.

I have procured about half a dozen specimens of this elegant little shell, some of which are in sufficiently good preservation to be identified. The opportunity of comparison with a recent specimen has been obligingly afforded me by Mr. Thompson of Belfast, and I have every reason to think they are the same. The apex is obtuse, but not reversed. There is a small but obscure fold upon the columella, scarcely discernible, except where the outer lip is broken away.

4. ODOSTOMIA SIMILLIMA (?). *Mont.* Tab. IX, fig. 6, *a—b.*

TURBO SIMILLIMUS. *Mont.* Test. Brit. Sup. p. 136, 1803.
—   *Wern.* Mem. vol. i, p. 406, t. 8, fig. 15.
—   *Turt.* Conch. Dict. p. 209, 1819.
TURRITELLA SIMILLIMA. *Flem.* Brit. An. p. 303, 1828.
RISSOA (?) COSTELLATA. *S. Wood.* Catalogue 1842.

Od. Testâ pusillâ, elongatâ, subulatâ, gracili; apice obtuso; anfractibus quinque convexis; suturis profundis; longitudinaliter costellatis; costellis rectis; aperturâ ovatâ, simplici, dimidiam spiram æquante; sine umbilico.

Shell minute, elongate, tapering, and slender, with an obtuse and reversed apex; volutions five, convex; suture deep, longitudinally costellated, with perfectly straight ribs; aperture ovate, rather elongate; last volution large, measuring half the axis.

Axis, $\frac{1}{8}$ of an inch.

Locality. Cor. Crag, Sutton.   Recent, British Seas.

I have but a few specimens of this elegant little shell, which appear to differ in some characters from the description of its recent analogue, to which I have referred it with doubt. It is much smaller, and the costæ are broad and smooth, having twenty in the last volution. The spaces between them are very narrow; the suture is sub-canaliculated, by the upper part of the outer lip being somewhat slightly raised, and there is a small obscure tooth upon the columella; my specimens are not in very good condition. Turton gives the recent shell as found on the shores of the island of Jura.

LITIOPA,* *Rang.* 1829.

BUCCINUM (spec.)   *Quoy et Gaim.*
PLANAXIS (spec.)   Id.
QUOYIA. *Deshayes.*
BOMBYXINUS. *Lesson.*

Gen. Char. Shell small, thin, corneous, or subhyaline, with a conical or tapering spire, and acute apex; volutions convex, smooth, or striated; aperture ovate, and rather large, with a sharp and simple outer lip, continuous to a truncated columella, and a subemarginate base. Operculum, none.

This genus was established by M. Sander Rang (in the Ann. des Sci. Nat., tom. xvi, p. 303), upon the little Pelagian species found in the Gulf-weed *Sargassum bacciferum*, and is placed by that naturalist in his systematic arrangement between Ianthina and Phasianella.

The shell, with its truncated columella, much resembles that of Achatina, and might perhaps be considered its marine representative. In the description by the above author, a "sommet pointu" is given as a generic character, which the recent species certainly possesses; but in the shell from the Crag, which I have presumed to belong to this genus, the apex is very obtuse. In my recent specimens, the first three volutions are contracted and shrunk, being covered with longitudinal wrinkles, as if that part of the shell might be dispensed with, without injury to the animal, like the upper volutions of the *Bulimus decollatus*, although I have never seen a specimen so decollated.

1. LITIOPA PAPILLOSA. *S. Wood.* Tab. IX, fig. 1, *a—b*.

   LITIOPA PAPILLOSA.   *S. Wood.* Catal. in Ann. and Mag. of Nat. Hist. 1842, p. 539, pl. 5, fig. 11.
   —   *Morris.* Catal. of Brit. Fossils, p. 149, 1843.

L. Testá ovatá, lævigatá, tenui, fragili; apice obtuso; anfractibus quatuor, convexiusculis; suturis profundis, subcanaliculatis; operturá ovatá; labro acuto, simplici; columellá truncatá.

Shell ovate, smooth, thin, and fragile, with an obtuse, rounded apex, and four slightly convex volutions: suture distinct and deep, or subcanaliculated; outer lip sharp, with a slightly replicate inner lip, forming a minute umbilicus, and a subemarginate aperture, or truncated columella.

Axis, $\frac{1}{6}$ of an inch.

Locality. Cor. Crag, Sutton.

My cabinet contains about half a dozen specimens of this pretty shell, which appears to be distinct from the only recent species I am acquainted with, and which is

---

* Etym. Λιτός, simple, and ὀπή, *foramen*, or orifice, as given by Hermansen. Or perhaps from λὶς, λιτός, a fine and delicate web, the animal spinning a thread by which it suspends itself to the floating weed.

peculiar to the Gulf weed. That species is covered with impressed striæ, and has a sharp-pointed apex, with more volutions than in the Crag shell. It is possible our shell may be an altered form, having lost its outer coating, and cast off its pointed apex, though its present appearance is much against its identity with the recent one. The recent shell has a more distinct truncation of the columella than is seen in my fossil specimens, where it is more like a plication, or folding over, of the columellar lip, behind which is the small umbilicus. Our shell has an ovate aperture, pointed at the junction of the outer lip with the body whorl, and slightly expanded at the lower part, with a papilliform and glossy apex, without, however, any remains of a roughened edge, which in all probability would have been the case had it ever possessed any greater number of volutions than are now visible.

SCALARIA,* *Lam.* 1801.
SCALA. *Humphries*, 1797.
SCALARUS. *Montf.* 1810.
ACIONA. *Leach*, 1815.
CLATHRUS. *Oken*, 1815.
— *Agassiz*, 1840.

Gen. Char. Shell turriculate, with an elevated spire, and generally acute apex, composed of rounded volutions, sometimes contiguous, at other times separated; smooth, transversely striated or decussated, longitudinally costated; costæ generally thin, erect, sharp, and somewhat oblique, often thick and obtuse; aperture entire, subcircular, rather larger than wide; peristome continuous, with a prominent and reflected margin; columella lip slightly everted, forming a subcanaliculated base; operculum thin, corneous, and spiral.

The shells composing this genus are of a vitreous texture, of a milky whiteness, semi-transparent, and are in general objects of great delicacy and beauty. They are common through all the tertiaries, and in the recent state are not restricted to any climate, being found in most parts of the world, in the Australian seas, the Indian ocean, and on the coast of Greenland. Some species are still included in this genus that do not possess a vitreous structure, but are opaque shells, and, when better known, may perhaps require to be removed.

* Etym. *Scala*, a ladder.

1. SCALARIA GRŒNLANDICA. *Chemn.* Tab. VIII, fig. 11 *a*, 11 *b*.

TURBO CLATHRUS GRŒNLANDICUS. *Chemn.* vol. xi, p. 155, t. 195, fig. 1878-79.
SCALARIA SIMILIS. *J. Sow.* Min. Conch. t. 16, 1812.
— SUBULATA. *Couthouy.* Bost. Journ. Nat. Hist. ii, p. 93, pl. 3, fig. 4.
— GRŒNLANDICA. *Gould.* Inv. of Massachus. p. 249, fig. 170, 1841.
— SIMILIS. *S. Wood.* Catalogue 1842.

Sc. Testá elongato-turritá, imperforatá; anfractibus ventricosis, contiguis, longitudinaliter costatis; costis obtusis, interstitiis transversim striatis, vel sulcatis; anfractu ultimo basi carinifero; aperturá ovato-circulari.

Shell elongato-turreted, thick, strong, and imperforate; volutions 9—10, ventricose and contiguous, longitudinally costated; costæ obtuse, oblique, transversely striated, or sulcated; base of last volution carinated; aperture subcircular.

Axis, 1⅛ inch.
Locality. Red Crag, Sutton.

Mam. Crag, Bramerton and Bridlington. Recent, Coast of Greenland.

This shell is rare in my cabinet as a Red Crag fossil, though, I believe, rather less so in the more recent formation of the Mammaliferous Crag. In my Catalogue it was considered distinct from the Greenland shell, but the examination of a greater number of specimens from Norfolk (in which a considerable degree of variation is exhibited), has induced me to alter that opinion.

2. SCALARIA VARICOSA. *Lam.* Tab. VIII, fig. 14.

SCALARIA VARICOSA. *Lam.* Hist. Nat. des An. sans Vert. 2d edit. tom. ix, p. 74, 1835.
— FIMBRIATA. Ency. Meth. pl. 451, fig. 4, *a*, *b*.
— — *S. Wood.* Catalogue 1842.

Sc. Testá subulatá, turritá, imperforatá; anfractibus contiguis, costatis; costis tenuibus, incumbentibus, crenato-fimbriatis, creberrimis; varicibus crassiusculis, irregulariter sparsis; transversim striatis, anfractu ultimo basi carinifero.

Shell tapering, turriculate, imperforate; whorls contiguous, suture deep, covered with numerous, thin, reflected, and fimbriated longitudinal costæ, and obtuse varices at irregular distances; transversely striated, having an obtuse elevated keel around the base.

Axis, 1¼ inch.
Locality. Cor. Crag, Sutton. Recent, Mediterranean.

This shell is at present rare in my cabinet. My specimens correspond precisely with a recent shell said by Mr. G. B. Sowerby, sen., to be from the Mediterranean, and which appears to be *Sc. fimbriata* of Lamarck.

The thickened varices sometimes form a continuous rib, running the whole way up the shell, but they are in general irregular both in number and position. The elevated transverse striæ are about ten in number; they are decussated by the small reflected and fimbriated costæ, which in the last volution are as many as 30 in number. The lower transverse band is elevated into a distinct keel around the base.

3. SCALARIA FIMBRIOSA. *S. Wood.* Tab. VIII, fig. 12.

TURBO LAMELLOSUS (?). *Broc.* Coq. foss. Subapenn. p. 379, t. 7, fig. 2, 1814.
Not. SCALARIA LAMELLOSA. *Lam.*
SCALARIA FIMBRIOSA. *S. Wood.* Catalogue 1842.

Sc. Testá crassá, subulatá, imperforatá, apice acuto; transversim sulcatá; longitudinaliter costatá; costis lamellosis, crenatis, crispis; anfractibus rotundatis, contiguis; basi carinatá; aperturá rotundá.

Shell thick, strong, tapering, and imperforate; apex acute; whorls round, contiguous, transversely sulcated, or striated; longitudinally ribbed, with thick reflected lamellæ, curved or fringed; base carinate; aperture circular; peristome reflected.

Axis, 1½ inch.

Locality. Cor. Crag, Ramsholt and Sutton.

This is a rare species, at least in my cabinet. The volutions are covered with prominent transverse or spiral ridges, between which there are finer striæ. A prominent obtuse keel runs round the base of the volution, and the peristome is reflected all round the aperture, particularly at the base of the columella. This shell is probably only a variety of *Turbo lamellosus*, Broc. (p. 379, t. 7, f. 2), although it does not appear to be so much fringed as the Italian fossil. I would nevertheless have given it Brocchi's name, but that it has been used by Lamarck for another species.

4. SCALARIA HAMULIFERA. *S. Wood,* 1847. Tab. VIII, fig. 13.

Sc. Testá subulatá, turritá, imperforatá; anfractibus, rotundatis; costellis lævibus, lamelliformibus, confertissimis, supernè aculeatis; interstitiis transversim striatis; anfractu ultimo basi vix carinifero.

Shell tapering and turriculate, with an imperforate base; volutions rounded, covered with numerous large, smooth, close-set lamelliform, and acutely-pointed or hook-shaped costæ, with an occasionally thickened subvariformed rib; striated or ridged spirally, with an obsolete keel upon the base.

Axis, ⅝ of an inch.

Locality. Cor. Crag, Sutton.

I have but one specimen of this shell, which, however, appears to differ so materially from any species I am acquainted with, that I feel disposed to consider it as distinct, and have given it a name for the present until the possession of a few more individuals will either confirm its correctness or remove it to one of its congeners. The same might almost be said of one or two others of this genus, that must for the present remain separated, as they cannot be connected by anything I possess. This shell presents the same differences to *S. fimbriata* and *S. fimbriosa* that *S. frondicula* does to *S. subulata*, the costæ being produced and acuminated at the upper part of the volution, but it has also larger, broader, smoother, and more numerously reflected costæ; they are unlike those of *S. fimbriata* and *S. fimbriosa*, which bear the curved impress of the spiral striæ or ridges, and in my single specimen the costæ are more oblique. In *S. fimbriosa* a continued line of rib passes from the apex to the outer edge of the last volution; whereas, in this species the line of rib is lost outside the penultimate whorl, although it is a shorter specimen. The ribs in the last volution amount to seventeen, with seven transverse striæ or ridges; the lower one around the base is larger than the others, but it scarcely deserves the name of a keel.

5. SCALARIA FRONDOSA. *J. Sow.* Tab. VIII, fig. 15.

SCALARIA FRONDOSA.   *J. Sow.* Min. Conch. t. 577, fig. 1, 1827.
                —   *S. Wood.* Catalogue 1842.

Sc. Testá elongato-conicá, lævigatá, imperforatá; anfractibus rotundatis, disjunctis costatis; costis tenuibus; supernè angulatis, acutis; aperturá rotundatá, marginatá.

Shell elongato-conical, smooth, imperforate; whorls round or cylindrical, disjoined; costæ thin, angulated, pointed, and projecting at the upper part; aperture circular; left lip recurved, subcanaliculated.

Axis, ⅞ of an inch.

Locality.   Cor. Crag, Sutton.

My cabinet contains but four specimens of this shell, which appears to be distinct. It approaches nearest to *S. foliacea*, particularly where the foliations are broken off; but I have never seen the elevated spine-like processes at the upper part of the lamellæ upon any of my numerous specimens of that species. It has about 8—9 volutions, with 9—10 lamellæ upon the last. The figure represents rather too many ribs upon one volution.

6. SCALARIA FRONDICULA. *S. Wood.* Tab. VIII, fig. 16.

SCALARIA FRONDICULA.   *S. Wood.* Catalogue 1842.
      —   FRONDOSA.   *Nyst.* Coq. foss. de Belg. p. 393, pl. 38, fig. 7, 1844.
      —   TENERA (?).   *J. Smith.* Quart. Journ. Geol. Soc. 1847, p. 421, fig. 24.

Sc. Testá elongatá, turritá; spirá subulatá; apice acuto; anfractibus rotundatis, contiguis costellatis; costellis lamellosis, supernè angulatis, spiniferis; interstitiis lævigatis; aperturá rotundatá.

Shell turreted and elongate; spire tapering; apex acute; whorls round, contiguous, with lamellated costæ, angulated, or rather spinous, at the upper part; whorls without striæ, and glossy; aperture circular.

Axis, ⅞ of an inch.

Locality.   Cor. Crag, Sutton and Gedgrave.

I have considered this as distinct from the preceding, in consequence of the volutions being contiguous; the lamellæ, also, are numerous, less prominent, and reflected, and it is a more subulate shell. It appears to resemble *S. clathratula*, particularly in the young state, but the costæ are elevated into small spinous fronds, projecting upwards around the upper part of the volution—a character I do not observe in my numerous specimens of that species, and the costæ are fewer in number (generally twelve), and thicker.

A Tertiary fossil, found in the beds of the Tagus, and recently described by Mr. Smith (in the Quarterly Journal of the Geological Society,) under the name of *Sc. tenera*, is perhaps the same species, its distinguishing character being a small mucronation at the upper part of each rib.

7. SCALARIA FOLIACEA. *J. Sow.* Tab. VIII, fig. 17.

> SCALARIA FOLIACEA. *J. Sow.* Min. Conch. t. 390, fig. 2, 1825.
> — *S. Wood.* Catalogue 1842.
> — *G. Sow.* Genera of Shells, fig. 5.

Sc. Testá turritá, acutá, imperforatá; anfractibus rotundatis, lævigatis, disjunctis costalis; costis 5—7, tenuibus, lamelliformibus, distantibus; aperturá rotundatá.

Shell turreted, spire elevated; apex acute; base imperforate, whorls cylindrical, smooth, and disjoined; ribs oblique, lamelliform, thin and distant, seven in the last volution; aperture circular, peristome reflected; without a keel upon the basal volution.

Axis, 1 inch.

Locality. Cor. Crag, Sutton.

Red Crag, Sutton.

An abundant shell in the Coralline, but very rare in the Red Crag. It differs from *Turbo pseudo scalaris*, Brocchi (t. 7, f. 1), which has a prominent line upon the base of the volution, though M. Philippi has given this shell as a synonyme to his *S. pseudo scalaris;* the ribs also are fewer in number, and the volutions more disjoined. The lamellæ form a continuous rib up the spire, in an oblique or slightly spiral direction, and are more especially reflected in the middle of the outer lip.

8. SCALARIA SUBULATA. *J. Sow.* Tab. VIII, fig. 18.

> SCALARIA SUBULATA. *J. Sow.* Min. Conch. t. 390, fig. 1, 1825.
> — *S. Wood.* Catalogue 1842.
> — *Nyst.* Coq. foss. de Belg. p. 394, pl. 38, fig. 8, 1844.

Sc. Testá turritá, elongatá, subulatá, lævigatá, imperforatá; anfractibus 10—11, rotundatis, disjunctis costellatis; costellis obtusis; in anfractu ultimo decem.

Shell turriculate, elongate, tapering, and smooth; whorls 10—11, rounded or cylindrical, disjoined and costated; ribs slightly oblique, ten in the last volution; base imperforate: aperture subovate.

Axis, ⅞ of an inch.

Locality. Cor. Crag, Sutton.

This species is not so abundant as *S. foliacea,* from which it differs in being more elongate, and in tapering gradually, with less fimbriated foliations. The volutions are perfectly smooth, differing in that respect from a recent species in my possession, which resembles it in all other respects, but that it is striated; the Crag shell does not appear ever to have the least vestige of striæ, and the peristome is much reflected and turned over, so as to form an obtuse varix, producing a nearly continuous rib up the spire.

9. SCALARIA CLATHRATULA. *Turt.* Tab. VIII, fig. 19, *a—b.*

TURBO CLATHRATULUS. *Turt.* Linn. vol. iv, p. 500, 1806.
　　　—　　　　　　*Turt.* Conchological Dictionary, p. 208, 1819.
TURBO PARVUS. *Maton* and *Racket.* Trans. Lin. Soc. vol. viii, p. 171, pl. 5, fig. 1, 1807.
SCALARIA MINUTA. *J. Sow.* Min. Conch. t. 390, fig. 3-4, 1825.
　　—　 PSEUDO-SCALARIS. *Dubois de Montp.* Foss. de Wolhyn. Pod. p. 43, pl. 11, fig. 36-37, 1831.
　　—　 PULCHELLA (?).　*Phil.* En. Moll. Sic. vol. i, t. 10, fig. 1, 1836.
　　—　 CLATHRATULA.　*S. Wood.* Catalogue 1842.

Sc. Testá turritá, lævigatá, imperforatá; anfractibus convexis, contiguis, costellatis; costellis confertissimis, lamelliformibus; aperturá subcirculari; marginibus acutis.

Shell turreted, smooth, and imperforate, with nine convex, contiguous volutions; costellæ rather variable in number, generally numerous, lamelliform, thin, depressed, and reflected; aperture subcircular, with a thin, reflected, sharp, and elevated margin.

Axis, $\frac{5}{8}$ of an inch.

Locality. Cor. Crag, Sutton.　　　　　　　　　　　　Recent, British Seas.

This species is exceedingly abundant and variable in the number of costæ; in some there are upwards of 20, while others have not more than 14 on one volution. It appears to differ from the *S. Trevelyana* principally in its more elongated form and less reflected costæ.　I have not yet seen this species from the Red Crag, although a strong shell, and very abundant in the older formation; in a recent state it is a very rare shell, and may probably be one of the species now dying out.

10. SCALARIA TREVELYANA. *Leach,* MS. Tab. VIII, fig. 20.

SCALARIA TREVELYANA. *Winch.* Geology of Lindisfarn, An. of Philosophy, xx, p. 434.
　　—　 TREVELIANA.　*Thorpe.* British Marine Conchology, p. 254, fig. 27, 1844.

Sc. Testá turritá, elongato-conicá, latiusculá, imperforatá, lævigatá; anfractibus convexis, subdisjunctis costellatis; costis tenuibus, reflexis, crebris, depressis.

Shell turreted, elongato-conical, imperforate, smooth; whorls convex, slightly detached; with numerous depressed costæ, thin and slightly reflected, about sixteen on the last volution; aperture suborbicular.

Axis, $\frac{3}{4}$ of an inch.

Locality. Red Crag, Sutton.　　　　　　　　　　　　Recent, British Seas.

I have found but one specimen of this shell, which appears to agree so well with a recent individual of *S. Trevelyana* in my possession, that I am induced to give a figure of it, although it is not perfect.　The last six species have the convoluted tube quite smooth, and free from lines or striæ of any kind.　Their specific separation depending principally upon the form and number of the reflections of the peristome, which generally produce a continuous line of rib along the spire, in a slightly oblique direction.　A larger number of specimens than I possess may possibly remove some of these distinctions.

11. SCALARIA (?) CANCELLATA. *Broc.* Tab. VIII, fig. 22, 22 *a*.
    TURBO CANCELLATUS. *Broc.* Coq. Subapenn. p. 377, t. 7, fig. 8, 1814.
    SCALARIA DECUSSATA. *S. Wood.* Catalogue 1842.

Sc. Testá subulatá, elongatá, tenui, fragili; spirá elevatá; apice acuto; longitudinaliter costatá; transversim striatá; anfractibus convexiusculis, rotundatis, marginatis: aperturá subrotundá.

Shell elongate, tapering, thin, and fragile, with an elevated spire and acute apex; whorls slightly convex, longitudinally costated, and transversely striated, reticulated; lower volution marginated, with the base striated; aperture subcircular, peristome continuous, sometimes thickened.

Axis, ½ an inch.

Locality. Cor. Crag, Sutton.

This species is delicate and fragile: my specimens, though numerous, are all more or less broken; the one figured is the most perfect; it is not more than half an inch in length, though fragments indicate a size at least half as long again.

12. SCALARIA (?) OBTUSICOSTATA. *S. Wood.* Tab. VIII, fig. 21.
    SCALARIA OBTUSICOSTATA. *S. Wood.* Catalogue 1842.

Sc. Testá turritá, subulatá, tenui, fragili; spirá elevatá; apice acuto; longitudinaliter costatá; costis obtusis: transversim tenuissimè striatis; anfractibus convexiusculis, marginatis; aperturá subrotundá.

Shell tapering; spire elevated, with an acute apex, thin, and fragile; longitudinally costated, and finely striated transversely; the striæ most conspicuous between the costæ; volutions slightly convex; base marginated, rather flat, and finely striated; mouth subcircular, peristome continuous.

Axis, ½ an inch.

Locality. Cor. Crag, Sutton.

Imperfect specimens of this shell are by no means rare; my largest does not exceed half an inch in length, while fragments indicate its full size to have been an inch. It somewhat resembles *Turbo lanceolatus*, Broc. (t. 7, f. 7), but appears to differ in having a greater number, as well as a greater convexity of the volutions; specimens half an inch in length have as many as 10—11 volutions, with about fifteen nearly erect and obtuse costæ in the last one; the outer edge of the volution is slightly marginated, so that the costæ terminate before reaching the suture. It is finely striated transversely, the striæ being most conspicuous between the costæ. This is a tender and fragile shell, with the surface slightly eroded.

The texture of this shell, as well as of the preceding species, is opaque and fragile, differing in composition from the true Scalariæ.

EULIMA,* *Risso*. 1826.

TURBO (spec.) *Linn.*
HELIX (spec.) *Mont.*
NISO. *Risso,* 1826.
PHASIANELLA. *Flem.* 1828.
PASITHEA. *Lea,* 1833.
BONELLIA. *Desh.* 1834.
POLYPHEMOPSIS (?). *Portlock.*

Gen. Char. Shell elongate, subulate, smooth, and glossy; spire elevated, composed of numerous volutions, sometimes contorted; apex generally acute; aperture ovate, rounded at the lower part, terminating in an acute angle above; outer lip simple, slightly obtuse, occasionally thickened, straight, or incurved; left lip slightly reflected; columella smooth; operculum corneous.

The glossy exterior, subulate form, generally small and almost undefined suture, with the flat volutions, will readily distinguish the shells of this genus from all others, when they are in good preservation. Species of this genus are considered to have existed in the earliest geological periods, but they are not at present well determined. Several new and beautiful species have been found by Mr. Edwards in the Eocene formations of England, and fifteen recent ones are figured by Mr. G. B. Sowerby, jun., in his ' Conchological Illustrations,' some of which have been placed in a new genus by Deshayes, under the name Bonellia.

1. EULIMA POLITA. *Linn.* Tab. XIX, fig. 1, *a—b.*

TURBO POLITUS. *Linn.* Syst. Nat. p. 1241, 1767.
— LÆVIS. *Penn.* Brit. Zool. t. 79, 1779.
— ALBUS. *Don.* Brit. Shells, pl. 177, 1802.
HELIX POLITA. *Mont.* Test. Brit. p. 398, 1803.
RISSOA BOSCII. *Payr.* Cat. des Moll. p. 112, pl. 5, fig. 15-16, 1826.
PHASIANELLA POLITA. *Flem.* Brit. An. p. 301, 1828.
MELANIA BOSCII. *Phil.* En. Moll. Sic. tom. i, p. 157, 1835.
EULIMA POLITA. - - - tom. ii, p. 134, 1844.
— ANGLICA. *G. Sow.* Conch. Illust. fig. 5, 1838.
— POLITA. *S. Wood.* Catalogue 1842.
— — *Thorpe.* Brit. Mar. Conch. p. 187, fig. 49, 1844.

Eu. Testá elongatá, lanceolatá, turritá, nitidissimá, politá; anfractibus tredecim, planiusculis, contiguis; suturá vix distinctá; aperturá ovatá, supernè acuminatá.

Shell elongate, lanceolate, subulate, smooth, strong, and glossy, with thirteen contiguous and flattish volutions; suture ill defined, scarcely distinct; outer lip nearly

* Etym. (?)

straight and obtuse; aperture subovate, acuminated at the upper part, with an inner lip slightly reflected.

Axis, ⅝ of an inch.

Locality.   Cor. Crag, Sutton.

Red Crag, Walton Naze.   Recent, British Seas and Mediterranean.

I have half a dozen well-preserved specimens, which present characters sufficient to justify this being considered identical with the recent British species.   Montague says, the volutions are quite flat; but in my specimens they are slightly convex in the recent as well as in the fossil shell, with a rather obtuse apex; outer lip not quite straight, projecting a little at the lower part, with sometimes an inflection in the spire (vide fig. 1 *b*).   Two of my specimens have five or six of the upper volutions broken off, and the opening closed in a manner similar to that of *Bulimus decollatus.*

2. EULIMA SUBULATA.   *Mont.* Tab. XIX, fig. 3.

HELIX SUBULATA.   *Mont.* Test. Brit. Sup. p. 142, 1808.

TURBO SUBULATA.   *Don.* Brit. Shells, t. 172, 1803.

HELIX SUBULATA.   *Broc.* Conch. foss. Subapenn. pl. 3, fig. 5, 1814.

MELANIA CAMBESSEDESII.   *Payr.* Cat. des Moll. p. 107, pl. 5, fig. 11-12.

—   *Phil.* En. Moll. Sic. tom. i, p. 157, 1836.

EULIMA LINEATA.   *Sow.* Conch. Illust. fig. 13, 1838.

MELANIA NITIDA.   *Grateloup.* Bord. Foss. p. 8, pl. 5, fig. 5, 1838.

EULIMA SUBULATA.   *Dujardin.* Mém. Soc. Géol. de France, t. 11, 1837.

—   *S. Wood.* Catalogue 1842.

Eu.   Testá elongatá, angustá, subulatá, acuminatá, politá, nitidissimá; anfractibus planatis, contiguis; aperturá elongato-ovatá, supernè acuminatá; labro simplici recto.

Shell elongate, tapering, smooth, and glossy, with an acuminated apex; whorls flat, contiguous; suture scarcely defined; aperture elongato-ovate, acuminated at the upper part; outer lip straight; inner lip slightly reflected.

Axis, ⅜ of an inch.

Locality.   Cor. Crag, Sutton and Ramsholt.   Recent, British Seas.

Small and mutilated specimens are by no means rare, and the above dimensions are those of my largest and most perfect ones.   There is no character I can detect, but that of size, in which it differs from the recent species, and mine are, in all probability, only young individuals.   My specimens do not possess more than eight volutions; recent individuals of the same length have about an equal number.   The apex in this species is sharper than in the preceding one.   A species very much resembling this shell, from the London clay, at Barton, is in the cabinet of Mr. Edwards.   It is rather more cylindrical and is less in size, with a more visible suture; his largest specimen does not exceed three eighths of an inch in its axis, and yet appears an adult shell. Although so close an approximation, I think it is distinct, and that gentleman is of the same opinion.

3. EULIMA GLABELLA. *S. Wood.* Tab. XIX, fig. 2.

EULIMA GLABELLA. *S. Wood.* Catalogue 1842.

Eu. Testá subulata, turritá, nitidissimá, politá; apice obtuso; anfractibus 6—7, planulatis, contiguis; suturis distinctis; aperturá ovatá, supernè acuminatá; labro arcuato.

Shell small, turreted, strong, naked, smooth, and glossy, with an obtuse apex; volutions 6—7, flattish; suture distinct; aperture ovate, acuminated at the upper part; outer lip rather obtuse and arcuated.

Axis, $\frac{3}{16}$ of an inch.

Locality. Cor. Crag, Sutton.

This is not a very rare species. It differs from *E. subulata* in having a more obtuse apex, a less elongated form, and in the curvature of the outer lip; the suture is better defined, and the whorls slightly convex. It may be known from the young and smaller specimens of *E. polita* by a more subulate form, and by the base of the body whorl not being angulated. It appears to be intermediate between the two. At the upper part of the volution, just below the suture, the shell is of a different tint, as if in its recent state it had been ornamented with a coloured band similar to that of *E. subulata,* only in a different position. It may possibly be the *intermedia* of Cantraine (Malac. Mediter. Mém. de l'Acad. Roy. de Brux., t. 13), but that shell is described as having a spire "effilée." Our figure represents the apex as too much acuminated.

ALVANIA,* *Leach,* MS. 1818.

ALVANIA. *Risso,* 1826.
ACLIS. *Lovén,* 1846.

Gen. Char. Shell elongate, subulate, and turriculate, with an elevated spire, and a papilliform and reversed apex; volutions numerous, convex, covered with elevated striæ; aperture ovate; outer lip thickened (?), with a slightly reflected inner lip, and a small umbilicus.

A species of this genus in the British Museum had long passed under the MS. name of *Alvania,* given to it by the late Dr. Leach (1818), and intended by him to have been erected into a genus, which name was afterwards adopted by M. Risso, 1826, with the following characters: "Coq. conique, suture presque profonde; ouverture ovale, aigué à droite; peretreme mince, parfait; opercule cornée." Although not very well defined, these characters are sufficient for the Crag shell it is intended here to include, and as the name originally given by Dr. Leach was attached to a specimen of *Alvania glabella,* the type of his genus, it is but right it should be retained.

* Etym. (?)

1. ALVANIA ASCARIS. *Turt.* Tab. XII, fig. 11, *a—c.*

TURBO ASCARIS. *Turt.* Conch. Dict. p. 217, No. 60, 1819.
TURRITELLA MINOR. *Brown.* Ill. Brit. Conch. pl. 51, fig. 57-58, 1827.
PYRAMIS LÆVIS. - - - pl. 50, fig. 51-52.
TURRITELLA ASCARIS. *Thorpe.* Brit. Mar. Conch. p. xlv, fig. 21, 1844.
ACLIS SUPRA-NITIDA. *Lovén.* Ind. Moll. Scand. p. 17, 1846.
ALVANIA SUPRA-NITIDA. *S. Wood.* Catalogue 1842.
— ALBELLA. Id. Catalogue, var. β, fig. 11 c.

A. Testá pusillá, turriculá, elongatá tenui; apice obtusiusculo; anfractibus 8—9, *convexis, striatis, vel sulcatis; sulcis circa quatuor, supra planatis; suturis profundis; aperturá ovatá, labro incrassato.*

Shell minute and slender, with an elevated spire, and slightly obtuse apex; volutions 8—9, convex, with three or four large, transverse, elevated, and rounded ridges upon each whorl; upper part plain and smooth; suture deep; aperture ovate, or subcircular, with a slightly curved and thickened outer lip; base of volution plain, with a slightly reflected left lip; umbilicus small.

Axis, $\frac{1}{7}$ of an inch.

Locality. Cor. Crag, Sutton. Recent, Coast of Ireland and North Seas.

About twenty specimens of this pretty little shell are in my cabinet, and as the upper volutions are smooth and glossy, and the upper part of the whorl is free from striæ, I considered it distinct from *ascaris;* for, in Dr. Turton's description, as well as in Brown's figure, the striæ are represented as regular and equidistant, and my dependence for comparison was upon their descriptions. Mr. Alder, however, has informed me that it is the same species as *Turbo ascaris,* Turt., and I have restored that name upon his authority. The first two or three of the volutions in all my specimens are without striæ, but that is probably from erosion. The apex of this shell is reversed and papilliform, and the external ridges are sometimes visible within the mouth, from the thinness of the shell, and two or three specimens have a slightly thickened varix on the outer lip. Two of my specimens have a diameter equal to nearly half the axis, while in others it is not more than a quarter. The dimensions of the aperture vary in length from a quarter to one third the length of its axis.

Alvania albella of my Catalogue is probably only an eroded form of this species. In comparing the Crag shells in the first instance with a specimen upon a tablet in the British Museum, having the name of *Alvania albella* in the handwriting of Dr. Leach, they were found to correspond sufficiently to be considered identical, confiding in the recent specimen as a perfect shell. As this was intended by Dr. Leach for the type of his genus, the name was used as such in my Catalogue, and continued here for the same reason. The introduction of numerous incorrect forms by Risso is not sufficient to invalidate the right of Dr. Leach to the above name.

Rissoa,* *Freminville* and *Desmarest*. 1814.

Turboella.　*Leach*, 1819.
Pyramis.　*Brown*, 1827.
Cingula.　*Flem*. 1828.
Cyclostrema.　Id.
Loxostoma.　*Bivona*, 1838.

Gen. Char. Shell small, elongato-ovate, or turriculate, with generally an elevated spire and acute apex, sometimes short and ovate, or subglobulous, smooth, striated, or costated; aperture suborbicular or ovate, acuminated at the upper part; peristome generally thickened and reflected, slightly intercepted at the base of the columella by an incipient notch; outer lip simple and sharp, or thickened and dentated within. Operculum corneous, closing the aperture.

This genus contains a large number of species, all of a small size, and shells are included, with different forms of aperture, which, when the animals are better known, may probably require to be separated; some have a circular, continuous, and thickened peristome, while others are subcanaliculated or slightly emarginate; many of them are strong and opaque shells, with thick and elevated costæ, while others are smooth, glossy, and vitreous, with a sharp and simple outer lip. The name of Rissoa was proposed in 1814 by the above authors, who described several species in the ' Bull. de la Soc. Phil. de Paris.' This genus appears first in the Great Oolite, four species from which formation are described in ' Min. Conch.' Many of the known recent species are inhabitants of shallow water.

M. d'Orbigny has separated those species which have an indication of a sinus at the upper as well as at the lower part of the aperture, and has united them into a genus, under the name Rissoina.

1. Rissoa striata. *Mont*. Tab. XI, fig. 1, *a—b*.

Turbo striatus.　*Mont*. Test. Brit. p. 312, 1803.
Cingula striata.　*Flem*. Brit. An. p. 307, 1828.
Rissoa communis.　*Forbes*. Malac. Monen. p. 17, 1838.
— striata.　*S. Wood*. Catalogue 1842.
Not. Rissoa striata.　*Phil*. En. Moll. Sic. vol. i, t. 10, fig. 8.
Cingula striata.　*Thorpe*. Brit. Mar. Conch. p. 178, fig. 99, 1844.

R. Testá elongatá, turritá, subcylindraceá, apice obtuso; anfractibus sex, convexis, transversim striatis, et longitudinaliter supernè costatis; aperturá ovatá, peristomate integro, labro obliquo, extus incrassato, intus lævigato.

Shell small, elongate, subcylindrical, with an obtuse apex; whorls six, convex, transversely striated, and longitudinally costated upon the upper part of the volution,

* Etym. After the name of M. Risso.

costæ obsolete upon the lower portion; aperture ovate, rather narrow or angulated at the upper part; peristome continuous, with a slightly thickened outer lip, oblique and projecting; umbilicus covered; suture deep and distinct.

Axis, ⅛ of an inch.

Locality. Cor. Crag,. Sutton. Recent, British Seas.

Not a very abundant species. From a careful comparison with the well-known recent species I have no doubt of its identity. The last whorl is slightly contracted, or rather depressed, and thrown a little out of its regular descending volution, giving the lower half of the shell a cylindrical form; the base of the inner lip is slightly reflected, like that in the genus Scalaria, and covers the umbilicus. In some specimens the ribs may be seen over the whole whorl, but in general only so at the upper half of it.

2. RISSOA ZETLANDICA. *Mont.* Tab. XI, fig. 7, *a—b.*

TURBO ZETLANDICA. *Mont.* Linn. Trans. xi, p. 194, t. 13, fig. 3.
CYCLOSTREMA ZETLANDICA. *Flem.* Brit. An. p. 312, 1827.
RISSOA ZETLANDICA. *S. Wood.* Catalogue 1842.
— *Morris.* Catal. of Brit. Foss. p. 161, 1843.
CYCLOSTREMA ZETLANDICA. *Thorpe.* Brit. Mar. Conch. p. 158, 1844.

R. Testá crassá, ovato-conoideá; anfractibus 5—6, *convexis, tumidis; cingulis transversis, elevatis, et grandis, circa quinque, longitudinaliter costatis, decussantibus, alveolatis; apertura rotundo-ovatá; labro intus lævigato; ultimo anfractu cingulum ad umbilicum ferente.*

Shell strong, ovato-conical; apex rather obtuse, with five convex volutions, covered with four or five large elevated striæ or ridges, which are decussated by longitudinal ribs; aperture subcircular; inside of outer lip without denticulations, and furnished with a thick or double ridge at the umbilicus.

Axis, 3/16 of an inch.

Locality. Cor. Crag, Sutton. Recent, Zetland.

Not a very abundant species. It appears to be identical with the shell from the North Seas. The spiral ridges are at regular distances, excepting those below the volution, where, after a wider space, there are two ridges, or rather a double ridge, which covers the immediate portion of the base around the umbilicus. On the upper part of the whorl there is a slight depression, giving a small shoulder to the volution. The longitudinal costæ are carried over the base, though they are less elevated there than upon the body of the shell; and the transverse ridges are more prominent than the longitudinal ones, which are tuberculated at the point of decussation. The thick or double ridge gives a crenulated groove at the base, and it is this character which best distinguishes the species.

3. RISSOA SEMICOSTATA. *Woodward.* Tab. XI, fig. 10.

TURBO SEMICOSTATUS (?).   *Mont.* Test. Brit. p. 326, t. 21, fig. 5, 1803.
          —   *Woodward.* Geol. of Norf. t. 3, fig. 19, 1833.
RISSOA SEMICOSTATA.   *S. Wood.* Catalogue 1842.
          —   *Morris.* Catal. Brit. Foss. p. 161, 1843.

R. Testá brevi, ovato-conicá; anfractibus quatuor vel quinque, convexis, tumidis; suturis profundis, longitudinaliter costulatis; costulis nnmerosis, in ultimo anfractu dimidiatis; transversim striatis; aperturá subrotundá; labro intus dentato.

Shell short, small, and conical, with four or five convex and tumid volutions; suture deep and distinct, longitudinally costated; costæ disappearing on the lower half of the volution, transversely striated; striæ visible between the ribs and over the base; left lip slightly reflected over a small umbilicus; outer lip thickened, and dentated within.

Axis, $\frac{1}{10}$ of an inch; *diameter,* $\frac{1}{12}$.

Locality.   Mam. Crag, Bramerton.   ? Recent, British Seas.

This is a rare shell in my cabinet, and Woodward speaks of it also as not being abundant.   It has about twenty ribs upon the last volution, and has an obtuse apex.   It is, I presume, the *T. semicostatus* of Woodward, as far as can be determined from his indifferent figure, but probably not the *T. semicostatus* of Montague.   I would suggest the name of *curticostata* for this species, if it be really distinct.   In my Crag specimens the costæ are visible upon the upper volutions.

4. RISSOA VITREA. *Mont.* Tab. XI, fig, 3.

TURBO VITREUS.   *Mont.* Test. Brit. p. 321, t. 12, fig. 3, 1803.
CINGULA VITREA.   *Flem.* Brit. An. p. 308, 1828.
RISSOA (?) VITREA.   *S. Wood.* Catalogue 1842.
CINGULA VITREA.   *Thorpe.* Brit. Mar. Conch. p. 182, 1844.

R. Testá turritá, elongatá, subcylindraceá, lævissimá, politá, vitreá, pellucidá, albá, tenui fragili; anfractibus 4—5, parum convexis; suturis profundis; apice obtusiusculo; aperturá ovatá; labro simplici.

Shell small, turriculate, subcylindrical, smooth, glossy, thin, white, and semitransparent, with four slightly convex volutions; suture distinct and deep; apex rather obtuse; aperture ovate, contracted at the upper part; peristome sharp, continuous; outer lip plain; inner lip covering the umbilicus.

Axis, $\frac{1}{8}$ of an inch.

Locality.   Cor. Crag, Sutton.   Recent, British Seas.

This elegant little fragile shell is at present rare, and appears perfectly identical with the recent British species.   It differs materially, in its vitreous texture, its glossy and subhyaline appearance, from the general character of the shells of this genus, which are usually opaque, and more or less externally ornamented.

5. Rissoa punctura (?). *Mont.* Tab. XI, fig. 4, *a—b.*

      Turbo punctura. *Mont.* Test. Brit. p. 320, t. 12, fig. 5, 1803.
      Rissoa concinna. *S. Wood.* Catalogue 1842.
      Cingula punctura. *Thorpe.* Brit. Mar. Conch. p. xliii, 1844.

    R. Testá elongato-ovatá, turritá ; apice acuminato ; anfractibus sex, convexis ; suturis profundis ; lineis elevatis transversis ; longitudinalibusque decussatis ; aperturá ovato-orbiculari ; umbilico parvo ; labro incrassato, marginato ; intus denticulato.

Shell elongato-ovate, turreted with an elevated spire, and about six very convex volutions ; suture deep, covered with 6—8 elevated transverse striæ or ridges, crossed by longitudinal lines, regularly decussating the exterior ; aperture subcircular, with a continuous peristome, behind which is a small umbilicus ; outer lip thickened and denticulated within.

    Axis, $\frac{1}{8}$ of an inch.

    Locality. Cor. Crag, Sutton.   Recent, British Seas.

This elegant little shell is not rare in the Coralline Crag, though the specimens are often much rubbed, and the external markings nearly obliterated. Some specimens show a thickened varix upon the penultimate whorl, as if the animal had occasionally grown beyond the ordinary size. The punctures between the cancellations are not visible in my specimens. Fleming considers *T. punctura* of Montague to be the young of *T. reticulata ;* our Crag shell, from its thickened outer lip, is, I have no doubt, a full-grown species.

6. Rissoa reticulata (?). *Mont.* Tab. XI, fig. 5, *a—b.*

      Turbo reticulatus. *Mont.* Test. Brit. p. 322, t. 21, fig, 1, 1803.
      Cingula reticulata. *Flem.* Brit. An. p. 306, 1828.
      Not. Rissoa reticulata. *Phil.* En. Moll. Sic. vol. i, t. 10, fig. 14, 1836.
      Rissoa reticulata. *S. Wood.* Catalogue 1842.

    R. Testá ovato-conicá, brevi, crassá ; anfractibus quinque, rotundatis ; longitudinaliter plicatis ; transversim crassi-striatis ; striis reticulatis, nodulosis ; aperturá subrotundá, labro extus incrassato, intus denticulato ; labio tenui subreflexo.

Shell small, ovato-conical, short, thick, and strong ; volutions five, convex, longitudinally plicated or costulated ; with thick and coarse transverse striæ decussating the exterior ; aperture subcircular, with a thickened outer lip, dented within ; inner lip thin, rather spreading, covering the umbilicus.

    Axis, $\frac{1}{10}$ of an inch ; *diameter,* $\frac{2}{3}$ the length.

    Locality. Cor. Crag, Sutton.   Recent, British Seas.

This shell has about six or seven large, coarse, and elevated transverse lines or ridges, which pass over the ribs, making the exterior decussated and nodulous ; the costæ are vertical, and they vary in number from eleven to sixteen in the last volution, becoming obsolete over the base. In the adult state, the mouth is thickened externally, with a large and prominent obtuse varix, and there are about eight denticulations within the outer lip.

This may possibly be *R. Montagui*, Payr. (p. 111, t. 5, f. 13-14), but our shell appears to be less in size, and less ventricose. The elevated transverse ridges give an apparent canal at the suture, and the outer lip is continuous with the inner one; at the base of the inner lip or columella the shell is slightly reflected. A shell somewhat resembling this species, but larger, is not very uncommon in the London clay, at Barton; it is thickened within the outer lip, but has not the denticulations characteristic of the Crag shell. Our figure represents the striæ as rather too fine and too numerous.

7. Rissoa pulchella (?). *Phil.* Tab. XI, fig, 9, *a—b*.
Rissoa pulchella. *Phil.* En. Moll. Sic. 1836, p. 155, t. 10, fig. 12.

R. Testá ovato-conicá, ventricosá, inflatá, crassá, subperforatá; anfractibus quinque, convexis, tumidis, longitudinaliter costulatis; costulis circiter quatuordecim, in anfractu ultimo dimidiatis; aperturá ovatá, spiram subæquante; labro incrassato, intus denticulato (?).

Shell ovato-conoidal, ventricose, thick, and strong, with five convex volutions, longitudinally costated; costæ about fourteen, disappearing on the lower half of the body whorl, with an ovate aperture, and a thickened outer lip, denticulated (?) within.

Axis, $\frac{1}{8}$ of an inch.

Locality. Red Crag, Sutton and Kesgrave.   ? Recent, Mediterranean.

My cabinet contains five specimens of what appears to be the same as the Mediterranean shell; they have the inflated form and numerous ribs of M. Philippi's figure, but there are the remains of striæ upon the base of our shell, and traces of denticulations within the aperture, which do not appear to exist in the former. It is larger and more ventricose than any of my specimens of *R. semicostata*, yet it is possible it may be only a variety. The costæ in our figure are rather too numerous, and the volutions scarcely tumid enough.

8. Rissoa confinis. *S. Wood.* Tab. XI, fig. 6, *a—b*.
Rissoa confinis. *S. Wood.* Catalogue 1842.

R. Testá ovato-conicá, subperforatá crassá; apice obtuso; anfractibus quinque, convexiusculis; suturis profundis, longitudinaliter costatis; costis rectiusculis, elevatis, in anfractu ultimo 9—12, transversim striatis, aut sulcatis; aperturá subrotundá; labro extus incrassato, intus dentato; labio vix reflexo, umbilico parvo.

Shell small, ovato-conoidal, perforate, thick and strong, with a rather obtuse apex; volutions five, convex; suture distinct and deep, longitudinally costated; costæ 9—12, nearly erect, elevated, covered with depressed transverse striæ; aperture suborbicular; outer lip thickened, dentated within; peristome continuous; the left lip slightly reflected, partly covering a small umbilicus.

Axis, $\frac{1}{10}$ of an inch.

Locality. Cor. Crag, Sutton.

This is an abundant species, and one I have not been able to identify with any known form. It approaches near to *Turbo parvus*, Mont. (p. 310), *Cingula parva*, Flem. (p. 306), where that species is incorrectly stated to be without spiral striæ. It appears to differ, however, in several particulars. Our shell has eleven or twelve ribs on the body whorl, and is covered with spiral striæ, or rather fine depressed sulci, which are carried over the ribs, and upon the base; it has a less elevated and a less acuminated spire than *R. parva*, and has also denticulations within the outer lip. Although abundant, the specimens of this species from the Crag are generally rubbed or eroded, particularly at the ribs, where a portion of the shell has disappeared, leaving a double ridge, with a depression between them. Occasionally, the shell may be seen quite perfect, with the striæ carried over the ribs; whereas, in my most perfect specimens of *R. parva* the striæ are only visible between them. Our shell more resembles *R. rufilabris*, Alder, but the upper whorls of that shell are smooth, so also is the back of the body whorl, and our shell has a more orbicular aperture. On account of these differences, it has been thought necessary to keep them separated. The artist has given rather too many ribs on a volution.

9. RISSOA OBSOLETA. *S. Wood.* Tab. XI, fig. 11, *a—b.*
RISSOA OBSOLETA. *S. Wood.* Catalogue 1842.

R. Testá ovato-conicá, perforatá crassá; anfractibus quinque, convexis, ventricosis, transversim striatis, striis obsoletis; suturis profundis; aperturá subrotundá; labro extus incrassato, marginato, intus dentato; peristomate integro; umbilico parvo.

Shell ovato-conical, ventricose, with a perforated base; whorls five or six, convex and tumid; suture deep; transversely striated; striæ obsolete; peristome continuous; aperture subcircular, with a thickened and marginated outer lip; and about five or six obtuse teeth within the mouth.

Axis, $\frac{1}{8}$ of an inch; *diameter,* $\frac{3}{5}$ of the axis.
Locality. Cor. Crag, Sutton.

This is a very abundant species; but I have met with it only from one locality, and the specimens are all more or less rubbed or decorticated. The striæ, when visible, are coarser, and most distinct upon the base of the shell; and in a few individuals they may be detected upon the upper volutions; but in general the shells are quit smooth. In its deep suture and inflated volutions it somewhat resembles *T. ventrosus*, Mont.; but it is a shorter species, and in its original condition was covered with spiral striæ. Many of my specimens are more tumid than even the figure represents, and a large prominent and thickened varix is left upon the outer lip. Some specimens of this species have, no doubt, become smooth by attrition; but the obsolete character of the striæ has arisen, in all probability, from a partial decomposition of the exterior. This may not be new, but I am not acquainted with any species with which it can be identified. The only markings I can perceive in any of my numerous specimens are the striæ in a spiral direction.

14

10. RISSOA CRASSI-STRIATA. *S. Wood.* Tab. XI, fig. 13, *a—b.*
RISSOA CRASSI-STRIATA. *S. Wood.* Catalogue 1842.

R. Testá turritá, elevatá, crassá ; anfractibus 5—6, convexis ; longitudinaliter costatis ; costis 10—12, rectis, obtusis; transversim lineatis, lineis paucis elevatis; suturis profundis; aperturá subrotundá ; labro extus incrassato, marginato ; intus denticulato.

Shell turriculate and elevated, thick and strong, with five or six convex volutions ; a deep and distinct suture, externally ornamented with 10—12 vertical and obtuse costæ ; transversely striated ; striæ or ridges few and elevated, carried over the ribs ; aperture subcircular, with a thickened and marginated outer lip, denticulated within the mouth.

Axis, $\frac{1}{8}$ of an inch ; *diameter,* nearly $\frac{1}{2}$ of the axis.

Locality. Cor. Crag, Sutton.

This is also an abundant species, and I have found it only at one locality. It somewhat resembles *R. sculpta,* Phil. (En. Moll. Sic. vol. ii, p. 132, t. 23, f. 21), but it appears to differ in several characters : the costæ are more prominent than those represented at the above reference ; it has also a thicker and more prominent varix upon the outer lip, in the adult state, a deeper and more distinct umbilicus, and it has six or eight denticulations within the outer lip, with the volutions more convex. In perfect specimens the striæ are coarse and elevated, about six or eight in the last volution ; and these are continued over the ribs, which are seldom more than ten in number. Many specimens have the costæ worn smooth, much resembling *R. obscura* of the same author (t. 23, f. 10).

11. RISSOA COSTULATA. *S. Wood.* Tab. XI, fig. 12, *a—b.*
RISSOA COSTULATA. *S. Wood.* Catalogue 1842.

R. Testá turritá, elongatá, perforatá; anfractibus 5—6, convexiusculis supernè subangulatis; longitudinaliter costulatis; costulis numerosis, confertis, rectis ; transversim striatis; suturis profundis; aperturá ovatá; labro simplici, intus denticulato ; umbilico parvo.

Shell elongate, turreted, and slender, with 5—6 slightly convex volutions, sub-angulated at the upper part, with a deep, well-defined suture ; costulæ numerous, 20—25, and erect, crossed by transverse impressed striæ, carried over the ribs ; aperture ovate ; peristome disconnected ; outer lip sharp, denticulated within ; left lip slightly reflected ; umbilicus small and open.

Axis, $\frac{1}{7}$ of an inch ; *diameter,* $\frac{2}{5}$ of the axis.

Locality. Cor. Crag, Sutton.

This shell is not very abundant ; I have a few specimens which are sufficiently perfect to exhibit well-marked characters. It somewhat resembles *T. decussatus,*

Mont. (t. 12, f. 4). The volution of our shell is obtusely but distinctly angulated on the upper part, giving the suture a deep depression. The specimens are generally more or less rubbed, causing the ribs to appear smooth; but one individual has distinct striæ carried over the ribs. I am not acquainted with any recent species with which it can be identified, and have therefore left it with its provisional name. Small shells found in sand, even in a recent state, have their characters sometimes obliterated, from friction by the agitation of the waters; in addition to which, many species from he Crag, particularly in this genus, have the exterior altered by decomposition or decortication, which renders their identification a matter of great difficulty.

12. RISSOA SUPRA-COSTATA. *S. Wood.* Tab. XI, fig. 8.
RISSOA SUPRA-COSTATA. *S. Wood.* Catalogue.

R. Testá elongato-conicá, turritá; apice parum obtuso; anfractibus 5—6, convexis; superioribus costulatis, et transversim striatis; anfractu ultimo lævigato; aperturá subrotundá; labro extus incrassato, intus denticulato; labio vix reflexo; umbilico parvo.

Shell elongato-conical, turreted, thick, and strong, with a somewhat obtuse apex; volutions 5—6, convex, suture deep; upper whorls longitudinally costated, and transversely striated; last whorl smooth and tumid; aperture subcircular, with a continuous peritreme; outer lip thickened and reflected, with 4—6 denticulations within; and a small but distinct umbilicus.

Axis, $\frac{1}{8}$ of an inch; *diameter,* $\frac{1}{2}$ the length of axis.

Locality. Cor. Crag, Sutton.

This is not a very abundant shell in my cabinet, and the specimens are rubbed, their true character being thereby somewhat obscured. It bears a great resemblance to *Pyramis similis* (Brown's Illust. Brit. Conch. pl. 51, f. 20), which M. Lovén considers as a var. of *membranacea*, Adams, *Helix labiosa*, Mont. As it is at present doubtful I have left it with its provisional name, although probably it may be a known species. The upper whorls are furnished with 10—12 rounded and vertical costæ, while the lower one is smooth and naked, and it has not the slightest vestige of a thickening or fold upon the columella. *Rissoa nana*, Grat. (Foss. de Bord.), has the upper volutions costated, but has no umbilicus.

The names of several of the species in this genus are given as new, while probably they are not so; being, however, unable to identify them with any known forms, they are left for the present with the provisional names given to them in my Catalogue.

PALUDESTRINA,* *D'Orb.* 1838.

SABANÆA. *Leach,* MS. 1818.
HYDROBIA. *Hartman,* 1821.
LITTORINELLA. *Brown,* 1843.
PALUDINELLA. *Pfeiffer,* 184— (?).
— *Lovén,* 1846.

Gen. Char. Shell generally small, of an ovate or ovato-conical form; spire more or less elevated; volutions convex and smooth, and in the recent state, of a corneous or semitransparent texture; aperture ovate; operculum corneous and spiral.

Most of the known recent species of this genus are estuary animals, and some are found where the water is occasionally fresh. They were separated from Paludina by M. d'Orbigny in consequence of a difference in the structure of the animal, which in this genus has the eyes at the anterior part of the base of the *tentacula*, whereas in Paludina the eyes are at the extremities of the tentacles. In the former the operculum is increased by spiral laminæ, but in the latter (Paludina) they are concentric. This genus nearly resembles that of Rissoa, the animals of which are, however, more strictly marine.

1. PALUDESTRINA SUBUMBILICATA. *Mont.* Tab. XI, fig. 2, *a—b.*

TURBO SUBUMBILICATUS. *Mont.* Test. Brit. p. 316, 1803.
— THERMALIS (?). *Linn.* Syst. Nat. p. 1237.
PALUDINA MURIATICA (?). *Lam.* Hist. des An. sans Vert. 2d edit. tom. viii, p. 463, 1838.
— *Dujardin.* Mém. Soc. Géol. de France, tom. ii, pt. 2, 1837.
TURBO MINUTA. *Woodward.* Geol. of Norf, t. 3, fig. 20, 1833.
RISSOA SUBUMBILICATA. *S. Wood.* Catalogue 1842.
— *Morris.* Catal. of Brit. Fossils, p. 161, 1843.

P. Testâ conicâ, turritâ, lævigatâ, subperforatâ; anfractibus quinque, convexis, tumidis; suturis profundis; apice obtusiusculo; aperturâ ovatâ, peritremâ integrâ; labro simplici, labio subreflexo, umbilico parvo.

Shell conical, elevated, smooth, and naked, with five convex and rather tumid volutions; suture deep and well defined; apex slightly obtuse; aperture ovate, with a continuous peritreme; inner lip a little reflected over a small umbilicus.

Axis, ⅛ of an inch; *diameter,* ½ the length of axis.

Locality. Mam. Crag, Bramerton. Recent, Britain.

This is an exceedingly abundant species in the estuary deposit at Bramerton, and I have one doubtful specimen from the Coralline Crag. This shell appears intermediate, between *P. ulvæ* and *P. ventrosa,* differing from the former in being more elongated, and in having the volutions more convex, and from the latter in having the same characters less prominent.

* Etym. *Paludina-æstuarina,* belonging to an estuary.

2. PALUDESTRINA ULVÆ. *Penn.*

Locality.   Cor. Crag, Gedgrave.   Recent, Britain.

A few specimens of this very common estuary species were found by myself on the upper part of the Coralline Crag, along with some *Helices, Cyrenæ,* and the usual common marine shells of that formation.   The spot whence they were obtained is near the surface, and it may possibly have been some disturbed and redeposited material of a more modern period.   My specimens are strongly tinged with their natural colour, and have the aspect of recent shells.

3. PALUDESTRINA (?) TEREBELLATA. *Nyst.* Tab. XII, fig. 7.
    MELANIA TEREBELLATA. *Nyst.* Coq. foss. de Belg. p. 413, pl. 38, fig. 12, 1844.

P. Testá turritá, elongato-conoideá, crassá, lævigatá (?) ; anfractibus septem vel octo, planulatis ; ultimo anfractu obtusé angulato ; aperturá ovatá utrinque subacutá ; labio reflexo, umbilico tecto.

Shell turreted, of an elongato-conical shape, thick and strong, with seven or eight rather flat and smooth (?) volutions ; last whorl subangulated at the base ; aperture slightly acuminated above and below ; left lip a little reflected, covering the umbilicus.

Axis, ⅝ of an inch.

Locality.   Red Crag, Sutton and Walton Naze.

This is not a very abundant species, and is always in a more or less mutilated and altered condition ; all my specimens have been much rubbed and decorticated, and the lines of growth are not clearly distinguishable.   Its appropriation to the present genus is a doubtful one.

4. PALUDESTRINA (?) PENDULA. *S. Wood.* Tab. XII, fig. 6.
    EULIMA PENDULA. *S. Wood.* Catalogue 1842.

P. Testá turritá, subulatá, elongatá, lævigatá ; anfractibus 8—9, convexiusculis ; aperturá ovatá ; labio subreflexo.

Shell turreted, tapering, elongated, and smooth, with eight or nine slightly convex volutions ; aperture ovate, acuminated at the upper part ; left lip slightly reflected.

Axis, ⅝ of an inch.

Locality.   Red Crag, Walton Naze.

Although not a very rare shell, my specimens are all much rubbed or decorticated; those which are most perfect have about eight slightly convex volutions, with an aperture resembling shells of this genus.   This may be also considered doubtfully appropriated to this genus.   It is an elegantly-formed shell, somewhat resembling an elongated ear-drop.

PALUDINA,* *Lamarck*.
HELIX (spec.) *Linn.*

Gen. Char. Shell ovato-conical, or subglobulous, and in the recent state thin, corneous, or semitransparent spire, more or less elevated, with rounded or convex volutions; aperture subcircular or ovate, slightly impressed by the previous whorl, and somewhat angulated at the upper part; operculum corneous or calcareous, formed of concentric laminæ.

Shells belonging to this genus in the recent state are found in water that is quite fresh, and the habits of the animal are much like those of Lymnææ, although it is said to be furnished with pectinated branchiæ.

This genus first appears in the Wealden formation, in what is considered by Dr. Mantell to be a fresh-water deposit, or probably the embouchure of an enormous river. In the Eocene formations of this country at Hordwell, a species of this genus is found associated with what may be considered as purely fresh-water animals, while the species found in the Mammaliferous Crag are mixed with decidedly marine species.

1. **PALUDINA LENTA.** *Brander.* Tab. XII, fig. 1, *a—c.*
　　　HELIX LENTA. *Brand.* Foss. Hanton. fig. 60, 1766.
　　　VIVIPARA LENTA. *J. Sow.* Min. Conch. t. 31, fig. 3, 1813.
　　　　　— CONCINNA. 　-　-　- 　t. 31, fig. 4-5.
　　　CYCLOSTOMA UNICOLOR. *Oliv.* Voy. pl. 31, f. 9 *a, b.*
　　　PALUDINA LENTA. *Desh.* Coq. foss. des Env. de Paris, p. 128, pl. 15, fig. 5-6, 1825.
　　　　　— SEMICARINATA. 　-　-　- 　p. 127, pl. 15, fig. 11-12.
　　　　　— MEDIA. *Woodward.* Geol. of Norf. t. 3, fig. 5-6.
　　　　　— ROTUNDATA. 　-　-　- 　t. 3, fig. 7, 1833.
　　　　　— OBSOLETA. 　-　-　- 　t. 3, fig. 4.
　　　　　— UNICOLOR. *Lam.* Hist. des An. sans Vert. 2d edit. tom. viii, p. 513, 1838.
　　　　　—　　— *S. Wood.* Catalogue 1842.
　　　　　—　　— *Cailliaud.* Voy. à Méroé, pl. 60, fig. 7.
　　　　　—　　— *Savigny.* Egypt, pl. 2, fig. 30.

B. *Testá ovatá, conoideá, crassá; apice acuto; spirá elevatá; anfractibus quinque, rotundatis, lævigatis; aperturá obovatá, supernè acuminatá, peristomate integro; umbilico obtecto.*

Shell ovato-conical, with an elevated spire and acute apex; volutions convex and smooth, with occasionally a few obsolete striæ; aperture subcircular, slightly acuminated at the upper part; margin continuous, and umbilicus closed.

Axis, 1 inch.

Locality. Mam. Crag, Bramerton and Thorpe.　　　　　　Recent, Nile River.

This shell is by no means rare at the above localities, and much too numerous to be considered as casually introduced, like *Planorbes, Lymnææ*, and the *Helices*. The specimens are variable in form, as may be seen in the figures referred to in our own plate, and I have no doubt the shells figured by Woodward belong to one and the

---

* *Palus-udis*, a marsh or lake; the general place of habitation for these animals.

same species. As a general character, the specimens from the Mammaliferous Crag have the volutions rather less rounded or convex than those from Hordwell, although there is considerable variation, in that respect, in the shells from both formations. A much greater variation of form in the volution, as well as in size and elevation, may be seen in *Valvata antiqua*, from the Fluviatile deposits of Grays and Clacton, than can be pointed out among those of this genus which I have considered identical, and that shell is now considered by most conchologists as only a variety of *V. piscinalis*. In the specimens of this species from the Crag the apex is generally more or less eroded, while it is rarely so in those I have procured at Hordwell, in which it is rather sharp than otherwise.

In examining some specimens from the Nile, in the British Museum, I could detect no differences by which they could be fairly considered as specifically distinct from the Hordwell species, and upon another tablet, in the same splendid collection, were specimens bearing the locality of China, which even more resembled the Hordwell shell; some of these have a subcarinated form of volution, with occasionally a few obsolete spiral striæ, and these same markings may be seen upon specimens I have myself collected at Hordwell; and the present shell I have presumed to be the prolonged existence of the British Eocene species. The animals of this genus, in the recent state, are now found inhabiting waters only that are quite fresh, while those from the Mammaliferous Crag are mixed with marine shells in greater numbers than could be supposed to be by an accidental introduction as dead specimens. The Paludinæ, therefore, of this period, must have either inhabited waters within the reach of the marine tide, or they must have been carried by the streams in large numbers to where they are now found. I am inclined to believe they lived very near to their present locality, as they bear no marks of transportation, and young specimens are by no means scarce.

Fig. 1 *a* is from the cabinet of Mr. Wigham.

2. PALUDINA TENTACULATA. *Linn.* Tab. XII, fig. 2.

HELIX TENTACULATA.	*Linn.* Syst. Nat. p. 1249, 1766.
—	*Mont.* Test. Brit. p. 389, 1803.
CYCLOSTOMA IMPURUM.	*Drap.* Moll. p. 36, pl. 1, fig. 19, 1805.
BITHINIA TENTACULATA.	*Gray.* Turt. Man. 2d edit. p. 93, pl. 10, fig. 120, 1840.
—	*S. Wood.* Catalogue 1842.
—	*Morris.* Cat. of Brit. Foss. p. 139, 1843.
PALUDINA IMPURA.	*Lam.* Hist. des An. sans Vert. 2d edit. tom. viii, p. 514, 1835.
—	*Rössmasler.* Icon. vol. i, p. 107, t. 2, fig. 65, 1836.

P. Testá ovato-conoideá, lævigatá, politá, pellucidá; anfractibus quinque; ultimo tumido, ventricoso; spirá elevatá; apice acuto; aperturá ovatá, supernè acuminatá.

Shell ovato-conoidal, smooth, and polished; whorls five, convex, the last large and ventricose; spire elevated; apex acute; with an ovate aperture, slightly impressed by the body whorl, and acuminated at the upper part.

Axis, $\frac{1}{2}$ an inch.

Locality. Mam. Crag, Bulcham, Bramerton.   Recent, Britain.

This species was found by Captain Alexander at Bulcham, and by Mr. Wigham near Norwich, but by no means abundantly at either locality. The specimen figured is from the cabinet of Mr. Lyell. There does not appear to be the slightest difference between the few specimens I have seen from the Mammaliferous Crag and those of the present day. This shell is found in great abundance in the truly Fluviatile deposits of Stutton and Clacton, and also its operculum. A slight variation in form exists among these specimens, some being much more ventricose than others, but the same differences may be observed in the recent shells. This species was separated from *Paludina* by Mr. J. E. Gray, and formed into a genus under the name of *Bithinia*, in consequence of its possessing a calcareous operculum, and from the shell being thickened internally behind the peritreme; these characters, however, can scarcely be considered as more than specific, as the animals are said to be precisely similar, and the different species, like the Naticæ, are furnished some with a calcareous, others with only a corneous operculum.

VALVATA, *Müller*, 1781.

NERITA (spec.) *Müller*.

TURBO (spec.) *Mont.*

Gen. Char. Shell conical, or sometimes discoidal, thin, corneous, or semitransparent, with rounded or cylindrical volutions, not impressed by the preceding whorl, generally smooth, sometimes carinated; aperture nearly circular, with a sharp and continuous peritreme, not reflected; operculum corneous.

This is a true fresh-water genus, and the few individuals found in the Crag associated with marine and estuary shells are, in all probability, transported specimens. Some shells of this genus approach so near in form to species of the genus Planorbis as t be almost undistinguishable in a fossil state. It may be remarked that the aperture in this shell is always more distinctly orbicular, and not at all impressed by the previous volution.

1. VALVATA PISCINALIS. *Müll.* Tab. XII, fig. 3.

NERITA PISCINALIS. *Müller.* Verm. p. 172, No. 358, 1774.

HELIX PISCINALIS. *Gmel.* Syst. p. 3627, 1788.

TURBO FONTINALIS. *Mont.* Test. Brit. p. 348, 1803.

—   *Don.* Brit. Shells, iii, t. 102.

CYCLOSTOMA OBTUSUM. *Drap.* Moll. t. 1, fig. 14, 1805.

VALVATA PISCINALIS. *Flem.* Brit. An. p. 286, 1828.

—   *Gray.* 2d edit. Turt. Man. pl. 10, fig. 114.

—   *Dujardin.* Mém. Géol. de France, tom. xi, pt. 2.

—   *S. Wood.* Catalogue 1842.

—   *Morris.* Catalogue of Brit. Foss. p. 166, 1843.

VALVATA ANTIQUA.   Id.

—   *G. Sowerby.* Mag. Nat. Hist. 1838, p. 574.

V. Testá tenui, globoso-conoideá, subtrochiformi, vel turbinoideá, perforatá; spirá mediocrè elevatá; anfractibus 4—5, convexis, rotundatis; suturá depressá; aperturá rotundá; umbilico profundo.

Shell thin, globosely conical, subtrochiform, or rather turbinated; spire slightly elevated; whorls 4—5, convex, rounded; suture deep; aperture circular; peritreme sharp, continuous; umbilicus moderate, deep.

Elevation, ¼ of an inch; *diameter of base,* the same.

Locality. Mam. Crag, Bramerton. Recent, Britain.

This shell I believe to be rare; two specimens of it were given to me by Mr. Lyell, and a few others sent for description by Mr. Wigham are all that I have seen. This species has also been found plentifully in the Fluviatile deposits of Grays, Clacton, and Stutton; and from the former locality a shell was described by Mr. Morris and Mr. Sowerby, under the name of *antiqua,* which was larger, more elevated, and with less convex or rounded volutions; and it was considered by those gentlemen a distinct species. Specimens have been since found possessing an intermediate character, some of which are in my own cabinet; and it is now generally considered by most conchologists only as a variety.

VERMETUS,* *Adanson,* 1757.
SERPULORBIS. *Sassi.*

Gen. Char. Shell elongate cylindrical, tubular, spiral in its young state, afterwards twisted without regularity; aperture small and circular, adherent generally by the apex or upper volutions; operculum corneous, circular, and multispiral.

This approaches so near to the testaceous Annelides, that it cannot be distinguished by the shell alone. By the examination of the animal, malacologists have, however, determined it to be a true Mollusc, approaching, in many of its characters, to the genus Turbo.

1. VERMETUS INTORTUS. *Lam.* Tab. XII, fig. 8, *a—b.*

SERPULA INTORTA. *Lam.*

VERMETUS INTORTUS. *Brown.* Lethea Geogn. p. 990, Taf. 36, fig. 18, *a, b, c,* 1837.

— *S. Wood.* Catalogue 1842.

— *Phil.* En. Mol. Sic. vol. ii, p. 144, 1844.

Dale. Hist. and Antiquities of Harwich, t. 13, fig. 1, 1730.

V. Testá plerumque gregariá, tereti, subquadratá, arctè spiratá, fere tota affixá, repente; striis transversis undulato-rugosis; lineis tribus longitudinaliter instructis, extremitate anticá aliquando liberá, porrectá.

Shell generally gregarious, tubular, subquadrate, laterally affixed for a considerable length, closely and regularly spiral in the young state, transversely rugose, with three

* Etym. *Vermis,* a worm.

longitudinal ridges, the centre one the largest; the extremity produced, straightened, drawn out, and free; aperture subcircular.

Diameter of tube, $\frac{1}{2}$ a line.

Locality.   Cor. Crag, Ramsholt and Sutton.

Red Crag, Sutton, Bromswell, and Brightwell.

Fragments of this shell are abundant in many localities.   One agglomerated mass in my cabinet is almost as large as a hen's egg; but the specimen figured by Dale, which I imagine to be a group of these tubes, is more than five inches in length, and three and a quarter in thickness.   It much resembles *V. subcancellatus,* Phil. (En. Moll. Sic.), but is not regularly cancellated on the exterior.   The tube is ornamented by three lines, in a longitudinal direction, the centre one of which is the most prominent, and forms the base or keel of the volution, so that only two are visible where the shell is regularly spiral, and one upon the base below.   The shell, when perfect, is covered with rugæ or wrinkles in a transverse direction; but the terminal portion is smooth, and free.   In large masses the volutions are irregular and inconspicuous, though the young shell may frequently be traced by its regularly spiral form.   *V. subcancellatus* is called by M. Philippi solitary; the Crag specimens are generally gregarious.   This may, however, be only a variety of that species, as he has suggested.

Fig. 8 *a* is a specimen attached to an oyster from the Coralline Crag; fig. 8 *b* is a glomerated mass of these shells, from the Red Crag at Bromswell.

2. Vermetus Bognoriensis (?). *J. Sow.* Tab. XII, fig. 9.

Vermetus Bognoriensis. *J. Sow.* Min. Conch. t. 596, fig. 1-3.

Specimens, of what appears to belong to this species, are occasionally found in the Red Crag.   The inner volutions are destroyed in a similar manner to those from Bognor, and the interior of the tube is filled with clay.   They are probably washed in from the London clay, and may, perhaps, be the remains of an Annelid.

Cæcum,* *Flem.* 1817.

Dentalium (spec.) *Mont.*

Cæcalium. *Flem.* 1822.

Brochus. *Brown,* 1827.

Cornuoides.   -   -

Odontina. *Zborzewsky,* 1834, (ex *Phil.*)

Odontidium. *Philippi,* 1836.

Dentaliopsis. *Clark,* MS. 1847.

Gen. Char. Shell small, tubular, generally thick and strong; smooth or annulated; subcylindrical and arcuated; open at the anterior extremity, with the margin rounded, sometimes thickened; closed posteriorly with a rounded or mammillated clausum; operculum corneous and spiral.

* Etym. *Cæcus,* blind, or closed; taken from *cæcum,* a term used in anatomy.

The animal of this genus has recently been examined by W. Clark, Esq., who considers it a highly organized Mollusc, and closely allied to the Trochidæ. Some notes upon the subject by that gentleman were obligingly sent to me for perusal by J. Gwyn Jeffreys, Esq., of Swansea, with permission to give the following extract from those remarks: "The Rev. M. J. Berkeley (in the Zoological Journal, vol. v, p. 426, at the end of the first paragraph) says, 'It is highly probable that the other minute British Dentalia will prove to possess an animal of like structure,' that is, of his new genus Ditrupa, which is not a Mollusc, but a genuine Annelidon. He is in error, as we have observed the animals of three species of minute Dentalia, viz. *D. imperforatum, trachea, glabrum*, and we have figured and described the first in our MS. They are all genuine Molluscs, much more highly organized than the Dentalia." He then gives copious anatomical details of the animal, which in a work devoted only to the calcareous remains of the Mollusca it is not necessary to repeat; and concludes by remarking, that he considers the animal of this genus to be a compound of Cerithium, Rissoa, Trochus, and of his new genus Pherusa, possessing a thick, corneous, circular, and multispiral operculum. Mr. Jeffreys further says there is little doubt of the accuracy of the foregoing remarks, and that he has himself verified those observations.

Dr. Fleming, in an article upon Conchology, published in the Edinburgh 'Encyclopædia, 1817,' proposed the name of Cæcum as a generic term for these little tubular and imperforate Testacea, which, in his work upon the 'Philosophy of Zoology,' he named Cæcalium; these shells were, however, afterwards transferred, in his 'History of British Animals, 1828,' into the genus Orthocera. Captain Brown, in his 'Illustrations of British Conchology, 1827,' gives figures of what he considered as seven different species, under the generic name of Brochus, and M. Philippi has described one species with the name of Odontidium; while, in the notes by Mr. Clark, these animals are described under the name of Dentaliopsis. As priority undoubtedly belongs to Dr. Fleming's name of Cæcum, I do not think it can be fairly superseded by even the excellent and important anatomical details of Mr. Clark.

A species from the Eocene formation at Hordwell, found by Mr. Edwards, is, as far as I know, the earliest appearance of this genus.

1. Cæcum trachea. *Mont.* Tab. XX, fig 5.

        Dentalium trachea. *Mont.* Test. Brit. p. 497, t. 14, fig. 10, 1803.
        Cæcum trachea. *Flem.* Edin. Ency. vol. vii, p. 67, 1817.
        Orthocera trachea. *Flem.* Brit. An. p. 237, 1828.
        Brochus trachiformis. *Brown.* Conch. Illust. pl. 1, fig. 10, 1827.
        Odontina (ex *Phil.*) *Zborzewsky.* 1834.
        Cæcum trachea. *S. Wood.* Catalogue 1842.
        Dentalium trachea. *Thorpe.* Brit. Mar. Conch. p. 5, fig. 61, 1844.
        Dentaliopsis trachea. *Clark.* MS. 1847.

C. Testá minutá, cylindricá, arcuatá, crassá, politá; transversim rugulosá; anticè apertá, margine annulatá; posticè obliquè truncatá, submucronatá.

Shell small, cylindrical, thick, strong, and glossy, transversely rugose; anterior extremity open, margin thickened or annulated; posterior end with an eccentric, obtusely-pointed clausum.

Length, ⅛ of an inch.

Locality.　Cor. Crag, Sutton.　　　　　　　　　　　　　　　　Recent, Britain.

This shell is not very abundant.　The closed termination is obliquely truncate and obtusely pointed, precisely resembling the recent species in that character, but my Crag specimens are not so deeply or regularly ridged or annulated; and in ten recent individuals that I possess there is no thickened margin at the anterior or open termi‑ nation, although they are quite as large, and appear to be full-grown specimens.　The greatly enlarged figure represents the posterior termination with its obtuse clausum.

2. CÆCUM MAMMILLATUM.　*S. Wood.*　Tab. XX, fig. 4, *a—b.*
　　　　CÆCUM MAMMILLUM.　*S. Wood.* Catalogue, 1842.
　　　　—　*var.* SUBULATUM, t. 20, fig. 4, *b.*

C. Testá tereti, arcuatá, subcylindricá, crassá, lævigatá, politá; anticè apertá, marginatá; posticè clausá; clauso mammillato, excentrico.

Shell tubular, curved, subcylindrical, strong, smooth, and glossy; anterior extre‑ mity open, annulated, or marginated; posterior extremity with an elevated, mammilli‑ form, projecting, excentric clausum.

Length, ⅛ of an inch.

Locality.　Cor. Crag, Sutton.

Although this species is exceedingly abundant in the Coralline Crag, I have as yet been able to obtain it only from one locality　From its perfectly smooth and glossy exterior, as well as from the elevated and papilliform character of its clausum, I have considered it to be distinct.　The shell is thickened a little behind the margin, which is somewhat thin and sharp, and one specimen has what appears to be the remains of coloured bands, as if the shell, when recent, was not quite white.　The mammillated and projecting portion of the closed termination is near the outer or curved side of the shell, and stands up prominently to a considerable height.　The general form of this species is nearly cylindrical, with the diameter of the closed end almost equal to that of the other, but some specimens are of a subulate or tapering shape, having the diameter of the open end more than twice that of the closed one, which induced me at one time to consider it a distinct species (fig. 4 *b*), but a slightly tapering form may be also observed in some of the specimens of *C. trachea.*　The thickened margin is, I imagine, the result of age, and not a specific character.

3. CÆCUM GLABRUM. *Mont.* Tab. XX, fig. 6.

DENTALIUM GLABRUM. *Mont.* Test. Brit. p. 497. 1803.
CÆCUM GLABRUM. *Flem.* Edin. Ency. vii, p. 67, t. 204, fig. 7, 1817.
ORTHOCERA GLABRA. *Flem.* Brit. An. p. 237, 1828.
DENTALIUM GLABRUM. *Turt.* Conch. Dict. p. 40, 1819.
BROCHUS GLABRUS. *Brown.* Illust. Brit. Conch. pl. 1, fig. 3, 1827.
— LÆVIS. - - - pl. 1, fig. 6.
CÆCUM GLABRUM. *S. Wood.* Catalogue 1842.
DENTALIUM GLABRUM. *Thorpe.* Brit. Mar. Conch. p. 4, fig. 5, 1844.
DENTALIOPSIS GLABRA. *Clark.* MS. 1847.

C. Testá pusillá, cylindricá, arcuatá, lævigatá, politá; anticè simplici, non marginatá; posticè clausá, obtusá, convexá, submarginatá.

Shell very small, cylindrical, curved, smooth, and glossy; anterior extremity simple, not marginated or thickened; posterior termination, with an obtuse convex clausum, submarginated.

Length, $\frac{1}{16}$ of an inch.

Locality. Cor. Crag, Sutton. Recent, Britain.

I have only two specimens of this shell, they, however, are so precisely similar to the recent species, that I have no hesitation in considering them identical. They are perfectly smooth, smaller than any of my specimens of *C. mammillatum*, and they are more particularly distinguished by having the posterior termination closed by a perfectly convex clausum, marginated by the edge of the shell.

4. CÆCUM (?) INCURVATUM (?). *Walker.* Tab. XX, fig. 7, *a—b.*

SERPULA INCURVATA. *Walker.* Test. Min. Rar. fig. 11, 1787.
VERMICULUM INCURVATUM. *Mont.* Test. Brit. p. 518, 1803.
SERPULA INCURVATA. *Turt.* Conch. Dict. p. 156, 1819.
CORNUOIDES MAJOR. *Brown.* Illust. Brit. Conch. pl. 1, fig. 49, 1827.
SERPULA (?) RECTA. *S. Wood.* Catalogue 1842.

C. Testá minimá, lævigatá, politá, tenui, fragili; anticè cylindraceá, apertá, non marginatá; posticè incurvá, convolutá.

Shell very minute, smooth, glossy, thin, and fragile; anterior portion cylindrical, with a simple but not marginated aperture; posterior portion incurved, with two volutions.

Length, nearly $\frac{1}{2}$ a line.

Locality. Cor. Crag, Sutton. ? Recent, Britain.

Three specimens of this minute shell, found by myself in the sand of the Coralline Crag, appear to resemble the figures given of the recent species.

Mr. Jeffreys informs me that he believes this to be only the young or immature state of *Cæcum glabrum,* and that decollation takes place at a certain period of its existence when the posterior termination is differently closed. Until further observations shall have confirmed the correctness of this opinion, it may be as well to allow it to remain as a distinct species, in proximity with what are, no doubt, its true relations.

Fig. 7 *b* is a young or perhaps broken specimen; the straight part of the tube in fig. **7** *a* is nearly cylindrical.

If *Serpula incrassata* of Walker be the young of *Cæcum glabrum*, it is probable the shell from the Crag may be the young state of *Cæcum mammillatum*.

LITTORINA,* *Ferussac,* 1821.

TURBO (spec.)   *Linn.*
TRICOLIA (spec.)   *Risso.*
NERITOIDES.   *Brown,* 1827.
LITORINA.   *Lovén,* 1846.
MELARAPHA.   *Megerle,* 1832 (fide *Gray*).

Gen. Char. Shell turbinate, thick, solid, and strong, ovate or globular; spire acuminated, consisting of few volutions; externally smooth or striated; aperture elliptical or subcircular; peritreme entire, sharp-edged; columella rather flattened; operculum corneous and spiral, with lateral nucleus.

Shells of this genus have a strong resemblance in form to those of Turbo, with which they have been generally placed.   A considerable difference exists in the animal inhabitants, and in the recent state they may be distinguished by the operculum, which in this genus is thin and corneous, while in Turbo it is calcareous, thick, and ponderous; in which genus, also, the shells are of a nacreous texture.   They may be distinguished from Phasianella, which they somewhat resemble, by the want of a polished and glossy exterior.

These animals, as the name implies, are frequenters of shallow water, near the shore, and are generally found in salt water.

As a genus, it has an extensive geographical range, being found in all latitudes; and a species is given by Mr. **J.** Sowerby as belonging to the Silurian system.

1. LITTORINA LITTOREA.   *Linn.* Tab. X, fig. 14, *a—k.*

TURBO LITTOREUS.   *Linn.* Syst. Nat. 12th edit. p. 1232, 1766.
——   ——   *Brown.* Illust. Brit. Conch. pl. 46, fig. 1-9, 1827.
——   ——   *J. Sow.* Min. Conch. t. 71, fig. 1, 1814.
——   RUDIS.   -   -   -   t. 71, fig. 2.
——   LITTOREUS.   *Smith.* Strata Identif. p. 5, pl. 2, fig. 3, 1816.
——   CARINATUS.   *Woodward.* Geol. of Norf. t. 3, fig. 11, 1833.
——   VENTRICOSUS.   -   -   -   - fig. 12.
——   BICARINATUS.   -   -   -   - fig. 13.
——   SULCATUS.   -   -   -   - fig. 14-15.
——   ELONGATUS.   -   -   -   - fig. 16-18.
DELPHINULA CARINATA.   -   -   -   - fig. 9.
LITTORINA VULGARIS.   *G. Sowerby.* Gen. of Shells, fig. 1.
——   SQUALIDA.   *Brod.* and *Sow.* Zool. Journ. vol. iv, p. 370, 1829.
——   ——   *Gray.* Zool. of Beechey's Voy. pl. 24, fig. 12.
——   LITTOREA.   *S. Wood.* Catalogue 1842.
——   ——   *Nyst.* Coq. foss. de Belg. p. 387, 1844.

L. Testá variabili, ovatá, imperforatá, lineis fuscis cinctá; apice acuto; anfractibus

* Etym. *Littus-oris,* the shore.

paucis, convexiusculis; transversim striatis, striis tenuibus, aliquando crassioribus, distantibus; anfractu ultimo ventricoso; aperturâ subovatâ; supernè acuminatâ; columellâ depressâ, obliquâ.

Shell very variable, more or less ovate, thick, strong, and imperforate, with an acute apex; spire generally depressed, but sometimes elevated; whorls slightly convex, transversely striated; striæ fine and small, sometimes coarse and elevated, often ornamented, with coloured bands; last whorl ventricose; columella depressed and oblique.

Axis, 1 inch.

Locality. Red Crag, Sutton, Bawdsey, &c.

Mam. Crag, Bramerton, Thorpe, and Bridlington.

Recent, British and North Seas.

This shell appears first in the Red Crag, where it is by no means abundant; but in the estuary deposit of the Mammiliferous Crag it may be found in great profusion. It is the most characteristic shell of that formation. This species may be considered as an arctic form, with a geographical range not very extended, while *Turbo rudis* is given by Dr. Gould as an American species. I have seen nothing among the numerous varieties from the Crag that could be fairly identified with the latter species, the shell considered to be so in 'Min. Conch.' being only a variety of *L. littorea.*

The principal difference in the specimens which I have considered as only varieties of one species appears to be in the lengthening or shortening of the convoluted cone, with an occasional thickening or elevation of some of the more prominent striæ. The form of the aperture is invariably the same, but it differs enormously in its proportions with regard to the axis of the shell.

In my Catalogue it is stated that the presumed cause of these variations was a more than ordinary alteration, and at various times, of the water in the estuary, which these animals are supposed to have inhabited; or perhaps the daily flow and ebb of the marine tide might for some time leave them in a purer fresh-water medium than was congenial to their nature, and this, probably, augmented occasionally by the introduction of floods of fresh water or ice. The animals of this species, with which the London markets are supplied, are obtained from the sea, and are more uniform in character.

Da Costa says, that the specimens of this species found on the shores of the Orkneys, are four times the size of those brought into the London markets. These shells are commonly called "Periwinkles," which is a corruption of petty winkle, or whelk. In Suffolk they are known by the name of "Pinpatches."

Fig. 14 is the normal or typical form of this species; from *a* to *k* are considered to be varieties, or mere monstrosities. Fig. 14 *d* is from the cabinet of Mr. Fitch.

2. LITTORINA (?) SUBOPERTA. *J. Sow.* Tab. X, fig. 13.

      VIVIPARA SUBOPERTA. *J. Sow.* Min. Conch. t. 31, fig. 6, 1813.

      LITTORINA SUBOPERTA. *Morris.* Cat. of Brit. Foss. p. 149.

      —   *Nyst.* Coq. foss. de Belg. p. 388, pl. 37, fig. 1, 1844.

L. Testá elonguto-conicá, crassá, rugosá; apice acuto; anfractibus angustis, convexiusculis; ultimo ad basim subangulato; aperturá ovatá; columellá callosá.

Shell elongato-conical, thick, and rugose, with an elevated spire, and acute apex; whorls depressed, elongated, subangulated at the base of the volution; aperture ovate, slightly contracted above and below; columella callous, left lip spreading over the umbilicus.

Axis, $\frac{3}{4}$ of an inch.

Locality. Red Crag, Sutton, Walton, Newbourn, and Bawdsey.

Specimens of this species may be found in numerous localities, but all which I have seen are in a mutilated and altered condition, having been more or less rubbed and broken. The shell is thick and strong, and generally rugose, except in those specimens that have been rubbed or eroded. The inner lip is rather extended, forming a small callosity, and covering the umbilicus.

FOSSARUS,* *Phil.* 1841.

PHASIANEMA. *S. Wood*, 1842.

Gen. Char. " Shell semiglobose, umbilicate, with an entire semicircular aperture, left or inner lip without folds (edentulous), never callous; umbilicus open; outer lip sharp, smooth within; operculum semiorbicular, corneous, not spiral."

Such are the characters given for a genus of shells by M. Philippi (in En. Moll. Sic. 1844, p. 147), and which he states to have been previously published in 'Archiv f. Naturgesch,' 1841. This is the revival of the name Fossar, used by Adanson, in his excellent work upon the Mollusca of Senegal in 1757, and which species M. Philippi considers as the type of the genus. When compiling my Catalogue, I was not aware of the existence of this genus; the name of Phasianema there proposed must of course give place to that by M. Philippi.

* Since the above has been in the press, I have discovered, in Mr. Gray's List of Genera, the name of *Forsar*, Gray, for this genus, with a date of 1840.

1. FOSSARUS SULCATUS. *S. Wood.* Tab. VIII, fig. 23, *a—d.*

>PHASIANEMA SULCATA. *S. Wood.* Catal. of Shells from the Crag, in An. and Mag. Nat. Hist. 1842, vol. ix, p. 935, pl. 5, fig. 15.

>— LINEOLATA. *S. Wood.* Catalogue, *var. β.*

>FOSSARUS CLATHRATUS. *Phil.* En. Moll. Sic. vol. ii, p. 148, t. 25, fig. 5, 1844.

Ph. Testá ovato-fusiformi; apice obtuso; anfractibus tribus, convexis; transversim sulcatis; interstitiis longitudinaliter plicatis; suturis depressis; labro acuto; columellá paululum uniplicatá.

Shell ovato-fusiform, with an obtuse apex; volutions three, convex; transversely sulcated, or rather ridged; longitudinally and finely plicated between the ridges; aperture large, ovate, acuminated at the upper part; outer lip sharp; inner one slightly replicate, with a small fold upon the columella.

Axis, $\frac{1}{7}$ of an inch.

Locality. Cor. Crag, Sutton.   Recent, Mediterranean.

Var. β, *lineolata* (fig. 23 *c, d*) was considered in my Catalogue as a distinct species, the sulci or ridges being greater in number, with a more tumid form in the volutions. The acquisition of more specimens has, however, removed some of these distinctions, and I am compelled now to include it only as a variety. About eight or ten elevated ridges cover the exterior of var. *a,* and these are sometimes at irregular distances, the furrows or spaces between them being wider than the ridges, and crossed by imbricated lines of growth; the shell is thin, and the ridges are visible from within.

This species is in all probability the same as M. Philippi has figured at the above reference, though his shell appears to be larger, with fewer elevated ridges. I have no specimen with less than eight, while some have as many as fifteen ridges, as in var β. In my specimens there is a thickening, or an obsolete tooth, about the middle of the columella, close to the umbilicus.

LACUNA, *Turton,* 1827.

>TEMANA. *Leach,* MS. 1819.

>TURBO (spec.) *Mont.*

>LUTEA., *Brown,* 1827.

>MACROMPHALUS. *S. Wood,* 1842.

Gen. Char. " Shell thin, conoid, or somewhat globular, clothed with an epidermis; aperture entire, rounded or oval, with the lips disunited at the top; pillar flattened, with a longitudinal groove, which terminates at the upper end in an umbilicus; operculum horny."

The above characters are given by Dr. Turton for this genus in the third volume of the 'Zoological Journal,' p. 190, where he also describes seven species, all of which appear to have a smooth exterior. Its principal distinction appears to be the elongated depression behind the edge of the inner margin of the aperture.

1. Lacuna reticulata. *S. Wood.* Tab. XII, fig. 10.

Macromphalus reticulatus. *S. Wood.* Cat. in An. and Mag. Nat. Hist. 1842, p. 537, pl. 5, fig. 16.

L. Testá minutá, fusiformi, tenui, fragili; spirá productá; apice subacuminato; anfractibus 4—5, convexis, longitudinaliter plicatis, plicis confertis, obliquis, et transversim striatis; decussantibus; aperturá ovatá; basi subacuminatá; labro acuto; umbilico magno lineari.

Shell small, fusiform, thin, and fragile, with an elevated spire, and subacuminated apex; volutions 4—5, convex; suture deep, longitudinally costated or plicated, with numerous oblique imbricated ridges, transversely striated, forming a reticulated surface; aperture ovate, lower part rather acuminated; peritreme sharp and continuous; umbilicus long and narrow.

Axis, ⅓ of an inch.

Locality. Cor. Crag, Sutton.

My cabinet contains only a few small specimens of a shell which I had considered in my Catalogue as generically distinct, in consequence of its elevated spire and reticulated surface. It is now restored to Turton's genus, until a greater number of better specimens shall determine its true position. The lines of growth are elevated into small obtuse costulæ. There are about twenty on the body whorl, which are crossed by elevated spiral lines. It much resembles the fusiform species of Cancellariæ or Admete, but it has a large and linear umbilicus.

At Tab. xv, fig. 12, the artist has inadvertently given a second figure of this species.

Turbo,* *Linn. Lam.* 1801.

Gen. Char. Shell turbinated, ovate, or conoidal, generally large, thick, and strong, with a short spire, and few volutions; aperture oval or subcircular; peritreme sharp and continuous, with a testaceous and solid operculum.

1. Turbo sphæroidea. *S. Wood.* Tab. XV, fig. 9.

Turbo (?) sphæroidea. *S. Wood.* Catal. of Shells from the Crag, in An. and Mag. of Nat. Hist. 1842, p. 533, pl. 5, fig. 3.

— Morris. Cat. of Brit. Fossils, p. 165.

T. Testá minutá, sphæroideá; spirá depressá; anfractibus tribus, convexis, rotundatis, striatis; striis 6—7, elevatis distantibus; peritremá acutá; umbilico magno aperto.

Shell very small, spheroidal, nacreous, with a slightly elevated spire; whorls three, convex, and rapidly enlarging; suture deep, spirally striated; peritreme sharp; umbilicus open, and surrounded by a prominent ridge.

Axis, half a line.

Locality. Cor. Crag, Sutton.

This pretty little shell is very rare in my cabinet. Its exterior is ornamented with about half a dozen elevated striæ, or rather ridges, placed at unequal distances upon

* Etym. *Turbo-inis,* a top or figure of an inverted cone.

the volution, becoming nearer together towards the lower part, and terminating with one that is elevated into a sort of keel, encircling the umbilicus, which is broad and naked, but not very deep.

TROCHUS,* *Linn.*

PYRAMIS. *Schum.* 1817.
ZIZIPHINUS. *Leach*, 1819.
GIBBULA. Id.
STEROMPHALA. Id.
PYRAMIDEA. *Swainson*, 1840.
CALLIOSTOMA. Id.

Gen. Char. Shell conical, with a more or less elevated spire, thick, strong, and of a nacreous texture; apex generally acute, sometimes depressed and obtuse; base flattish, with a carinated edge to the volutions; aperture square or rhomboidal, with an oblique columella, parallel to the outer lip; operculum corneous and spiral.

This is a marine genus, and not restricted to any climate, being found in all parts of the world. It is also of great antiquity; it appears first in the Protozoic rocks. Mr. G. B. Sowerby, in his genera, has separated Trochus from Turbo by the distinguishing character of a corneous operculum for the former, and a testaceous one for the latter; these two genera merge so imperceptibly into each other, that they are not to be defined by the form of the shell alone, although the species of this genus are generally more truly conical.

SECT. *a.* IMPERFORATE.

1. TROCHUS CRENULARIS. *Lam.* Tab. XIII, fig. 7, *a—b.*

TROCHUS CRENULARIS. *Desh.* Coq. foss. des Env. de Par. p. 229, pl. 27, fig. 3, and pl. 28, fig. 13, 15.
— *Bowdich.* Elements of Conch. pl. 9, fig. 7, 1822.

Tr. Testá elongato-conicá, pyramidatá, elevatá, imperforatá; transversim tuberculatá; anfractuum margine inferiore crasso, tuberculis majoribus crenato; columellá truncatá, incrasssatá.

Shell pyramidal or elongato-conical, imperforate; sides rather flat, ornamented with transverse tuberculated striæ, inferior one most prominent; base flat; aperture trapeziform, with a truncated columella, thickened, or unidentated at the lower part.

Axis, 1 inch.

Locality. Cor. Crag, Gedgrave.

The specimen from which the figure was taken is the only one I have seen; it was obligingly sent to me by Mr. Charlesworth for description: it is from the cabinet of H. Davey, Esq., of Beccles, and was purchased, among other Crag fossils, at Orford, in Suffolk. This specimen so precisely resembles some of my Grignon shells, that it is requisite to record the manner it has come into my possession, as there is a slight

* Etym. *Trochus,* a top.

doubt respecting the locality, and unfortunately there is nothing within the shell that will assist in the determination, though I am strongly inclined to believe it belongs to the Coralline Crag period, and that the shell was found at Gedgrave.

The lower part of the volution, near the outer angle, is furnished with a double row of tubercles, above which are two spiral threads, also tuberculated; the whole is crossed with distinct, oblique, subimbricated lines of growth, and upon the base are four or five circular or rather spiral ridges, also crossed by prominent lines of growth. It has a prominent tooth or projection at the lower part of the columella. The shell is beautifully nacreous beneath the outer coating.

2. Trochus ziziphinus. *Linn.* Tab. XIII, fig. 9, *a—h.*

> Trochus ziziphinus.   *Linn.* Syst. Nat. p. 1231.
> —   —   *Mont.* Test. Brit. p. 274, 1803.
> —   —   *Brown.* Illust. Brit. Conch. pl. 45, fig. 16-22.
> —   lævigatus.   *J. Sow.* Min. Conch. t. 181, fig. 1, 1817.
> —   concavus.   -   -   - t. 272, fig. 1, 1823.
> —   Sedgwickii.   *J. Sow.* Syst. Catalogue 1835.
> —   —   *Nyst.* Coq. foss. de Belg. p. 380, pl. 35, fig. 20, 1844.
> —   lævigatus.   -   -   - p. 379, pl. 36, fig. 11.
> —   conulus, *var.* β.   *Phil.* En. Moll. Sic. vol. i, p. 175, 1836.
> —   conuloides.   *Lam.* 2d edit. tom. ix, p. 142, 1843.
> —   ziziphinus.   *S. Wood.* Catalogue 1842.
> —   pseudo-ziziphinus.   Id.

Tr. Testá orbiculato-conicá, obliquá; apice acuto, granulato; anfractibus planis aut concavis; transversim et tenuiter striatis; ad marginem subcarinatis; infimá facie convexá; columellá obliquá; aperturá dilatatá, tetragoná.

Shell orbiculato-conical, oblique, fragile, with an acute apex and granulated spire; sides flat or slightly concave, covered with fine, or occasionally with coarse striæ; base of volution submarginate, and slightly extended, somewhat convex beneath, and concentrically striated; aperture trapezoidal.

Axis, 1½ inch; *diameter,* variable.

Locality.   Cor. Crag, Sutton and Ramsholt.

       Red Crag, Sutton.   Recent, Britain.

This handsome shell is not rare, but it is difficult to obtain it in perfect condition, from its extreme fragility. Nearly all my specimens are from the Coralline Crag; they are covered with fine striæ, and differ from the common form of the recent *T. ziziphinus,* which are more or less ridged. Prof. E. Forbes informs me he has dredged up specimens in the Scottish seas with the same kind of fine striæ, and as that is the only essential difference between our shell and *T. ziziphinus,* they are presumed to be identical. The first three or four volutions are covered with about five granular ridges, but in many specimens the upper ridge is the only one on which these granulations are preserved. The variety with concave sides has naturally a corresponding convexity of base, and the edge of the volution slightly extended, giving it a greater diameter.

Specimens from the Red Crag appear to be very rare; fig. 9 *h* is one that approaches nearer our common form; fig. 9 *f, g* is from the Coralline Crag, and appears to resemble *Tr. conuloides*, Lamarck. *Trochus lævigatus*, Sow., is probably only a specimen of this species, similar to fig. 9 *c*, which I imagine is an abraded individual from the Red Crag, upon which there is not the slightest vestige of striæ, the outer coating having been entirely removed; its only markings are the lines of growth; fig. 4 is also presumed to be a monstrous and decorticated specimen of this species.

3. TROCHUS CONULUS. *Linn.* Tab. XIII, fig. 5, *a—b.*

TROCHUS CONULUS.	*Linn.* Gmel. p. 3579.
—	*Lam.* 2d edit. Hist. Nat. des An. sans Vert. t. 9, p. 142, 1835.
—	*Phil.* var. *a.* En. Moll. Sic. vol. i. p. 175, 1836.
—	*Thorpe.* Brit. Mar. Conch. p. 255, 1844.
—	*S. Wood.* Catalogue 1842.

Tr. Testá conicá, nitidá, lævigatá; apice acuto, granulato; anfractibus planis, ad basim angulatis; infernè cingulo, crassiusculo marginatis, infimá facie planatá; aperturá tetragoná.

Shell conical, smooth, and naked, with an acute and granulated apex; sides flat, and a distinct ridge at the carinated edge of the volution; base nearly flat; columella suboblique; aperture trapeziform.

Altitude, 1 inch; *diameter,* $\frac{7}{8}$.

Locality. Cor. Crag, Ramsholt. Recent, Mediterranean and Britain.

But two or three specimens of this species have come into my possession, which appear to correspond with the smooth glossy form of what has hitherto been called *T. conulus*. It is very possible this may be a variety of *T. ziziphinus.*

Prof. Forbes and Mr. Hanley are of opinion that *T. conulus* is specifically distinct from *T. ziziphinus*, and as those gentlemen are better acquainted with the recent species than myself, I have given it upon their authority.

4. TROCHUS FORMOSUS. *Forbes.* Tab. XIII, fig. 2, *a—b.*

TROCHUS FORMOSUS.	*Forbes.* An. and Mag. Nat. Hist. 1847, vol. xix, p. 96, pl. 9, fig. 1.
— QUADRICINCTUS.	*S. Wood.* Catalogue 1842.

Tr. Testá conicá, imperforatá, granulatá; anfractibus quinque, planiusculis; cingulis quatuor, elevatis instructis; ultimo anfractu ad basim subangulato; aperturá tetragoná.

Shell regularly conical, with flat sides, and slightly elevated spire; volutions five, ornamented with about four elevated and granulated ridges; base rather flat; edge of volution subangulated, with a subquadrate or trapeziform aperture.

Axis, $\frac{1}{2}$ an inch.

Locality. Cor. Crag, Sutton.

Red Crag, Sutton. Recent, British Seas.

I have but a few specimens of this from either formation, and Prof. Forbes considers them identical with his recent British species. As the name given in my Catalogue was without description, although of prior date, it must of course give place to the one he has employed. The ridges, which are but four upon my specimens, are nearly equidistant, and the shell is smooth between them, with the exception of lines of growth; the lower ridge at the angle of the volution appears to be double. On those from the Coralline Crag, the upper ridge only is crenulated on the last whorl, the shell having lost a portion of its outer covering. In one from the Red Crag, the upper three ridges are distinctly nodulous, while the lower one alone is smooth. In the young state three volutions are crenulated upon all four ridges, which appear to vanish as the shell grows older, and it is beautifully cancellated on the obtuse apex. The recent analogue appears to have been obtained, living at the depth of fifty fathoms, by Mr. M'Andrew, about forty miles west of the main land of Zetland.

5. TROCHUS PAPILLOSUS (?). *Da Costa.* Tab. XIII, fig. 6, *a—c.*

     TROCHUS PAPILLOSUS.   *Da Costa.* Hist. Nat. Test. Brit. p. 38, pl. 3, fig. 5-6, 1778.
     —   GRANULATUS.   *Born.* Test. Mus. Vind. t. 12, fig. 9-10, 1780.
     —   PAPILLOSUS.   *Don.* Brit. Shells, 4, t. 127.
     —   TENUIS.   *Mont.* Test. Brit. p. 275, pl. 10, fig. 3, 1803.
     —   SIMILIS (?).   *J. Sow.* Min. Conch. t. 181, fig. 2, 1817.
     —   GRANOSUS.   *S. Wood.* Catalogue 1842.
     —   —   *Nyst.* Coq. foss. de Belg. p. 377, pl. 35, fig. 19, 1844.
     —   DEKINII (?).   -   -   -   -   pl. 36, fig. 10.

Tr. Testá orbiculato-conicá, obliquá, imperforatá, basi dilatatá; apice peracuto; anfractibus convexisculis; striis transversis, alternatim majoribus et granulosis; infimá facie planulatá, concentricè striatá et granulatá; aperturá dilatatá, tetragoná.

Shell orbiculato-conical, rather depressed, with a sharp and acute apex; volutions slightly convex, covered with transverse, granulate striæ; base somewhat flattened, imperforate, and concentrically striated; aperture expanded and trapezoidal.

Axis, ¾ of an inch.

Locality. Red Crag, Sutton, Newbourn, Bawdsey, Walton.   Recent, Britain.

Small specimens of this shell are abundant in several localities, particularly at Walton-on-the-Naze. From close examination of many specimens, I am induced to believe this to be only a variety of the British species *T. papillosus,* although the base is flatter, and the volutions a little more convex than in the generality of full-grown specimens of the recent shell. I imagine *Trochus similis* of Sow. to be a variety of this species, with flatter sides (fig. 6 *c*). The middle part of the base is generally free from concentric striæ, but in some specimens it is entirely covered, and the striæ on the upper surface alternate one large and one small.

6. TROCHUS SUBEXCAVATUS. *S. Wood.* Tab. XIII, fig. 8, *a—c.*

     TROCHUS SUBEXCAVATUS.   *S. Wood.* Catalogue 1842.

Tr. Testá conicá, elevatá, granulatá, imperforatá; anfractibus septem, supernè concavis subexcavatis, infernè convexis; striis transversis granulatis; basi planulatá, concentricè striatá et granulatá; aperturá tetragoná.

Shell conical and elevated, with seven volutions; upper part slightly concave; lower part convex, covered with about seven transverse granular ridges, and one small intermediate thread; base rather flat, concentrically striated; aperture trapezoidal.

Axis, 1 inch.

Locality. Red Crag, Sutton, Newbourn, and Walton Naze.

This species is by no means rare. It is presumed to be distinct from the preceding, from its being much more elevated, and from its having the upper half of the volutions deeply indented or concave. The variety fig. 8 *c* resembles *T. papillosus,* but is more elevated. It is only a different form of this species, if it be distinct; it is, however, possible they may be varieties of *T. papillosus,* but among my numerous specimens there are none that will fairly unite the two.

7. TROCHUS MILLEGRANUS. *Phil.* Tab. XIII, fig. 1, *a—b.*

TROCHUS MILLEGRANUS. *Phil.* En. Moll. Sic. vol. i, p. 183, t. 10, fig. 25, 1836.
— — - - - vol. ii, p. 154, 1844.
— — *Thorpe.* Brit. Mar. Conch. p. 264, fig. 36, 1844.
— MARTINI. *Smith.* Wern. Trans. vol. viii, p. 99, t. 1, fig. 26, 1838.
— ELEGANS. *Jeffreys.* MS.

Tr. Testá conicá, elevatá; anfractibus planatis, subimbricatis; basi marginatis, prominulis; ultimo anfractu angulato; transversim lineatis, lineis elevatis granulatis; basi planiusculá, striatá; aperturá rhomboideá.

Shell conical, elevated, thick, and strong; whorls flat, transversely striated, and granulated, with a carinated and marginated edge; base rather flat and striated; aperture rhomboidal.

Axis, $\frac{5}{8}$ of an inch.

Locality. Cor. Crag.   Recent, Mediterranean and Britain.

By a comparison with a recent specimen from the Mediterranean, given to me by Prof. Forbes, the Crag shell appears to agree with the recent species in most of its characters, although it has lost a good deal of its granulated exterior; the transverse ridges on the lower part are nearly smooth. It has a prominent line at the lower part of the volution, and a very slight thickening at the base of the columella. Prof. Forbes, who has seen my specimens, agrees with me in thinking that it is only an altered form of *T. millegranus.*

8. TROCHUS MULTIGRANUS. *S. Wood.* Tab. XIII, fig. 3, *a—b.*

TROCHUS ASPERULUS. *S. Wood.* Catalogue 1842.

Tr. Testá conicá, elevatá, crassá; anfractibus 5—6, planiusculis granulatis; basi marginatis, prominulis; ultimo anfractu subangulato; transversim lineatis, lineis elevatis granulatis, non imbricatis; basi planiusculá; columellá obliquá, obsoletè unidentatá.

Shell conical, elevated, thick, and strong; volutions 5—6, rather flat, ornamented with 6—8 transverse granulated ridges; last whorl subangulated; base somewhat flat, and a thickened or obscurely-toothed columella; aperture trapezoidal.

Axis, ¼ of an inch; *diameter,* ⅜.

Locality. Red Crag, Sutton.

This species is at present rare in my cabinet. It appears to differ from *T. mille-granus,* the species to which it approaches nearest, in having the ridges granulated; but it is without the imbricated character of that species, and the edges of the volutions are not so sharp and angular as in that shell, or in *T. miliaris;* the volutions, also, are not so flat. It differs from any of the specimens of *T. exiguus* that I have seen, in having the outer edge of the volutions less angulated, and in wanting the distinct basal ridge of that shell, and moreover its sides are less flat. Our shell has two or sometimes three prominent granular ridges at the lower part of the volution. It is more finely and regularly granulated than the young of *T. subexcavatus,* for which it might otherwise be mistaken.

Sect. b.   UMBILICATED.

9. Trochus villicus (?). *Phil.* Tab. XIV, fig. 4, *a—b.*
Trochus villicus.   *Phil.* En. Moll. Sic. vol. ii, p. 152, t. 25, fig. 14, 1844.

Tr. Testá conicá, obtusiusculá, crassá; anfractibus quinque, planulatis, supernè sub-angulatis; transversim cingulatis, cingulis 5—7, inæqualibus; anfractu ultimo angulato; columellá obsoletè unidentatá; umbilico parvo.

Shell conical, somewhat obtuse, thick, and strong, with about five volutions, rather flat, and projecting slightly at the upper part, ornamented with 5—7 unequal ridges, imbricated by the elevated and reflected lines of growth; base of volution angulated and striated; columella with an obsolete tooth; umbilicus very small.

Axis, ¼ of an inch.

Locality. Cor. Crag, Sutton.
       Red Crag, Sutton.   Recent, Mediterranean.

I have about half a dozen specimens of this shell in my cabinet from the Red Crag, which appear to resemble what M. Philippi has considered a new species; and in his examination of the recent shell he may have detected more decided characters for its establishment; but, judging from my fossils alone, I consider it only a variety of *T. Adansoni.* My specimens have but a very small umbilicus, which may arise from the position of the volutions being a little more erect; in other respects it corresponds so closely with his figure and description that it may be considered the same. From five to seven large and obtuse irregular-sized ridges cover the upper side of the volution, and there are about the same number on the base, but rather smaller and more regular; the ridges are all crossed by erect and slightly reflected lines of growth, which are most distinct between the ridges.

10. TROCHUS MONTACUTI. *W. Wood.* Tab. XIV, fig. 1, *a—b.*

TROCHUS MONTACUTI. *W. Wood.* Ind. Test. Sup. t. 6, fig. 43. 1828.
— LEUCOPHEUS (?). *Phil.* En. Moll. Sic. vol. i, t. 10, fig. 17, 1836.
— TURGIDULUS. *Dubois de Montp.* Wolhyn. Podol. p. 40, pl. 2, fig. 29-30.
— MONTACUTI. *S. Wood.* Catalogue 1842.
— — *Thorpe.* Brit. Mar. Conch. p. 255, fig. 18, 1844.

Tr. Testá conicá, elevatá, obtusiusculá, crassá ; anfractibus contiguis, planiusculis, ultimo anfractu obtusè angulato ; transversim striatis ; umbilico clauso, interdum perforato ; aperturá subquadratá ; columellá obsoletè unidentatá ; labro acuto, obliquo.

Shell conical and elevated ; spire slightly obtuse, thick, and strong ; whorls five, flat, transversely striated ; base of volution obtusely angular ; umbilicus generally covered ; aperture subquadrate ; columella with an obsolete tooth ; outer lip sharp and oblique ; lines of growth visible, slightly raised.

Axis, ¼ of an inch.

Locality. Cor. Crag, Sutton.

Red Crag, Sutton.   Recent, Britain.

This is by no means a rare shell in the Coralline Crag ; but I have met with only one specimen from the Red Crag. The volutions are rather flatter, and the base more angular, in the recent specimens which I have seen ; with the exception of these slight differences, they perfectly correspond, and the Crag shell may, I think, be considered as identical with the recent species. When quite perfect it has slightly reflected lines of growth, and bears a great resemblance to some varieties of *T. Adansoni,* and among my fossils they can scarcely be separated. Remains of the colour which originally ornamented the shell are visible on one of my specimens, crossing the lines of growth at an angle of about 45°.

11. TROCHUS ADANSONI. *Payr.* Tab. XIV, fig. 3, *a—c.*

TROCHUS ADANSONI. *Payr.* Catal. des Moll. de l'Ile de Corse, pl. 6, fig. 7-8, 1826.
— — *Phil.* En. Moll. Sic. vol. i, p. 182, 1836.
— ADRIATICUS. - - - vol. ii, p. 153, t. 25, fig. 10, 1844.
— OCTOSULCATUS. *Nyst.* Coq. foss. de Belg. p. 381, pl. 38, fig. 1, 1844.

Tr. Testá convexo-conicá, obtusiusculá, crassá ; anfractibus planiusculis, transversim striatis, vel cingulatis, crenulatis ; ultimo anfractu obtusè angulato ; columellá obsoletè unidentatá ; umbilico tecto, aliquando perforato.

Shell convexo-conical, thick, and strong, with rather flat sides, striated or ridged transversely ; striæ crenulated or imbricated ; base of the last volution obtusely carinate ; aperture subquadrate, with a sharp outer lip, and an obsolete tooth upon the columella ; umbilicus generally covered, sometimes open.

Axis, ¼ of an inch.

Locality. Cor. Crag, Sutton.

Red Crag, Sutton.   Recent, Mediterranean.

17

This is a very abundant shell in the Coralline Crag.   Some specimens are strongly marked by the remains of colour, which cross the lines of growth at a considerable angle, others resemble one of M. Philippi's varieties, "maculis longitudinalibus magis minusve confluentibus." The lines of growth are very distinct, and slightly elevated upon the striæ, and also between them; and the number of transverse ridges vary from six to ten.   The umbilicus is generally covered, but sometimes it is open.   A recent specimen from the Ægean Sea, obligingly sent to me for comparison by Professor E. Forbes, appears to correspond in all essential characters with my fossils, excepting that it is larger; and the Crag shells are rather more imbricated than the recent shell, which might naturally be expected from the loss of the epidermis.   The shell is beautifully nacreous beneath its outer covering.

12. Trochus Kicksii. *Nyst.* Tab. XIV, fig. 5, *a—b*.

Trochus Kicksii.   *Nyst.* Coq. foss. de Belg. p. 381, pl. 38, fig. 2, 1844.
—   Robinsii.   -   -   -   p. 382, pl. 38, fig. 3.

Tr. Testá convexo-conicá, obtusiusculá; anfractibus 5—6, convexiusculis, transversim tenuissimè striatis; striis impressis; columellá obliquá obsoletè unidentatá.

Shell convex or obtusely conical, thick, and strong; volutions 5—6, slightly convex, transversely striated, with five impressed striæ; aperture subrhomboidal; columella oblique, obsoletely toothed.

Axis, ¼ of an inch.
Locality.   Cor. Crag, Sutton.
                   Red Crag, Sutton.

A very abundant shell in the Coralline Crag.   The characters which most distinguish this shell are the impressed striæ and rather smooth exterior; it differs but in a slight degree from *T. Adansoni,* and is probably a variety of that species.   On the young shell more especially the striæ are but few and deeply impressed.   The form of the shell, and its slightly convex and overhanging volution, resembles the figure by M. Nyst; but his description, "undique lævigata," does not fully accord with our shell, as even in my most worn and altered specimens traces of striæ may be detected.

13. Trochus tumidus. *Mont.* Tab. XIV, fig. 2, *a—b*.

Trochus tumidus.   *Mont.* Test. Brit. p. 280, t. 10, fig. 4, 1803.
Gibbula tumida.   *Leach.* MS. 1819.
Trochus nitens.   *Woodward.* Geol. of Norf. t. 3, fig. 10, 1833.
—   littoralis.   *Brown.* Conch. Illust. of Brit. pl. 45, fig. 1, 4, 1827.
—   tumidus.   *S. Wood.* Catalogue 1842.
—   —   *Morris.* Catal. of Brit. Foss. p. 165, 1843.

Tr. Testá convexo-conicá, obtusiusculá, crassá; anfractibus tumidis, convexiusculis; suturis profundis; supernè subangulatis, transversim tenuissimè striatis; ad basim obtusè angulatis; aperturá subquadratá; umbilico mediocri, sæpe clauso.

Shell obtusely or convexo-conical, strong, and variable; whorls tumid, slightly convex, projecting or subangulated at the upper part, and obtusely angular at the base, finely striated transversely; aperture subquadrate, with a moderately sized umbilicus, which is sometimes entirely closed.

Axis, nearly ½ an inch.

Locality. Red Crag, Sutton.

Mam. Crag, Bramerton. Recent, Britain.

I have met with this shell at one locality only in the Red Crag, where it is not rare. In some specimens the spire is highly elevated, in others it is much depressed, while others, again, preserve an intermediate character. The umbilicus in my specimens is also variable; in some it is open, while in others it is quite covered, depending upon the elevation or depression of the spire, and also on the extension of the left lip: in the fossil that covering is sometimes broken away, but at other times it is open, where there has been no fracture.

14. TROCHUS CINERARIUS. *Linn.* Tab. XIV, fig. 7.

TROCHUS CINERARIUS. Syst. Nat. p. 1229.

— *Mont.* Test. Brit. p. 284.

TROCHUS LINEATUS. *Da Costa.* Brit. Conch. t. 3, fig. 66.

Tr. Testá crassá, orbiculato-convexá, vel obtuso-conicá; apice obtuso; anfractibus subplanulatis, transversim striatis; ad basim obtusicarinatis; aperturá tetragoná; umbilico tecto.

Shell thick and strong, obtusely or depressedly conical, with an obtuse apex, transversely striated; volutions rather flat, obtusely carinate at the edge; base somewhat flat; columella oblique; aperture subtrapezoidal; umbilicus closed.

Axis, ½ an inch.

Locality. Red Crag, Walton Naze and Sutton. Recent, Britain.

But few specimens of this species are in my cabinet. They appear to resemble the well-known recent species, and it is, in consequence, considered identical; a few remains of colour in one specimen, like that upon the recent shell, cross the lines of growth at nearly right angles, and the umbilicus is quite closed. *T. perforatus*, Smith (Mem. Wern. Nat. Hist. Soc. vol. viii, fig. 3-4), is probably a variety of this species.

15. TROCHUS CINEROIDES. *S. Wood.* Tab. XIV, fig. 8, *a—b.*

TROCHUS CINEROIDES. *S. Wood.* Catalogue 1842.

Tr. Testá crassá, orbiculari, depresso-conicá; apice obtuso; anfractibus convexiusculis, transversim striatis; striis crenulatis; umbilico magno aperto; columellá obtusè uniplicatá.

Shell thick and strong, obtusely conical; volutions slightly convex, transversely striated, with rough crenulated striæ; base of volution obtusely carinate, with a large open umbilicus, and a projection, or obsolete tooth, upon the columella.

Axis, ⅜ of an inch.

Locality. Red Crag, Walton Naze and Sutton.

This shell was published in my Catalogue as a distinct species, in consequence of its crenulated striæ, in which character it was considered by several eminent conchologists, as well as by myself, to be well marked. For this reason I have retained its present name, as also from its having a large and open umbilicus, by the possession of which it differs from the preceding species, which has it entirely closed. I cannot, however, but feel doubtful of the correctness of its separation, and that the possession of more specimens may probably unite the two. It somewhat resembles *T. umbilicatus*, Mont., in its large and open umbilicus at all ages, if it be a distinct species; but that shell is never crenulated like ours. In the young state it is much depressed, with a large and open umbilicus, the size of which, in the adult shell, depends upon the greater or less elevation of the spire, but it is always open. The spiral lines of striæ are about ten in number, generally alternating, one large and one small; all are more or less crenulated, and the peculiar depression in the umbilicus gives a tooth-like projection, or twisting, to the lower part of the columella. The circular ridges upon the base, in some specimens, are also strongly crenulated; these crenulations are probably more prominent and apparent in the fossil state than they would have been in the recent shell.

16. TROCHUS TRICARINIFERUS. *S. Wood.* Tab. XIV, fig. 6, *a—b.*
TROCHUS TRICARINIFERUS. *S. Wood.* Catalogue 1842.

Tr. Testá orbiculato-conoideá, crassá; apice acuto; anfractibus quinque, crenutatis; cingulis tribus, elevatis, instructis; suturis profundis; aperturá subquadratá; umbilico parvo; columellá obsoletè unidentatá.

Shell orbiculato-conoidal, thick, and strong, with a slightly elevated spire and acute apex; suture deep; whorls five, ornamented with three elevated obtuse carinæ, crenulated by elevated and reflected lines of growth; aperture subtrapezoidal, with an obsolete tooth upon the columella.

Axis, ¼ of an inch.

Locality. Cor. Crag, Sutton.

This shell, small specimens of which are not rare, somewhat resembles the figure and description of *Tr. Guttadauri*, Phil. En. Moll. Sic. t. 11, f. 1, but our shell has an umbilicus nearly closed, and the volutions are not rounded; the aperture is of a subquadrate form. Small specimens have the carinæ sharp, but in the older ones they become rounded and obtuse, and there is sometimes a small intermediate thread. The base has about half a dozen elevated striæ or ridges, and the lines of growth are elevated into fine reflected fimbriæ, most conspicuous in young shells and between the ridges. The upper keel is much elevated, leaving a deep sulcus round the top of the whorl. The outer lip is sharp, sloping inwardly to the thick part of the shell.

The imbricated lines of growth on some specimens are close and numerous, on others they are fewer in number, and more distant. Whether M. Philippi's description of "sulci foveolati" relates to the spaces between these imbrications I cannot say. There is nothing in our shell like the holes represented in the enlarged portion of his figure: I have therefore left it for the present with the name given to it in my Catalogue, until it can be correctly determined by a comparison with the Mediterranean specimens.

17. TROCHUS OBCONICUS. *S. Wood.* Tab. XIV, fig. 10, *a—d.*
TROCHUS OBCONICUS. *S. Wood.* Catalogue 1842.

Tr. Testá pusillá, orbiculato-conoideá, perforatá ; apice obtuso ; anfractibus quatuor, supernè planulatis, subtus convexis ; transversim lineatis, lineis elevatis, crenulatis ; suturis profundis, subcanaliculatis ; umbilico parvo.

Shell small, obconical ; apex rather obtuse ; whorls somewhat flat above, convex beneath, with transverse crenulated lines or striæ; suture deep, subcanaliculated; aperture obtusely quadrate, and a small umbilicus.

Axis, ¹/₇ of an inch.

Locality. Cor. Crag, Sutton.

This is an abundant shell, and my specimens are all from one locality. In general the transverse lines are regular in distance, and are about eight in number, crossed by elevated and imbricated lines of growth, which give the exterior a crenulated appearance ; the lines being prominent, particularly the upper and lower ones, produce an apparent canal at the suture. In some specimens there are only four elevated lines, var. β, fig. *a—b.* The base is convex, with about eight spiral lines ; there is a small umbilicus, and the shell is nacreous. It is possible this may be *Tr. crispulus,* Phil., but my specimens have not the rounded volutions of that shell as represented in his figure, (t. 25, f. 12.)

18. TROCHUS DITROPIS. *S. Wood.* Tab. XIV, fig. 9, *a—b.*
TROCHUS BICARINIFERUS. *S. Wood.* Catalogue 1842.

Tr. Testá minutá, orbiculato-conicá, perforatá; spirá depressá; anfractibus 3—4, rotundatis, bicariniferis; transversim striatis, et tenuissimè imbricatis; aperturá orbiculato-quadratá; umbilico parvo profundo.

Shell small, orbiculato-conical, striated along the volutions, and very finely imbricated across them ; whorls round, convex beneath, with two sharp and prominent keels ; aperture subquadrate ; umbilicus rather small, and deep.

Axis, ¹/₁₀ of an inch.

Locality. Cor. Crag, Sutton.

This pretty little shell is not very abundant, and I have as yet seen it but from one locality. The volution bears two elevated and sharp carinæ, dividing it into three unequal parts ; the spaces above and beneath being wider than that between the keels ; but, as the volution joins the lower keel, one carina only is visible upon the spire. The transverse or spiral striæ are broad and flat ; about half a dozen are on the upper half of the volution, and the same number are on the base, with only three between the keels; those upon the base are most prominent and distinct.

In Savigny's great work upon Egypt is the representation of a shell somewhat resembling this species, but which, from the figure, appears to be too strongly imbricated, and to have the aperture too circular, to be considered identical; a comparison with the shell itself is necessary for such determination.

Many of the shells in this genus also have the exterior much altered by decortication, and the determination of some of the species is not by any means satisfactory.

MARGARITA,* *Leach*, MS. 1819.

TROCHUS (spec.) *Linn.*

SOLARIELLA, *S. Wood.* 1842.

Gen. Char. Shell thin, subconical, trochiform; spire moderately elevated; volutions few, rather inflated; striated, sometimes costated; aperture subcircular; peritreme sharp, discontinuous, interrupted by the previous volution; umbilicus generally large and deep; operculum corneous, multispiral.

The animal of this genus much resembles that of Trochus, and it is considered by some conchologists as not deserving of generic distinction. There is, however, a delicacy and generally a semitransparency in these shells, accompanied by a large and open umbilicus in all the species, that at least entitles them to a sectional separation, and as the name was employed in my Catalogue, I have retained it here.

Mr. G. B. Sowerby, sen., has described fifteen recent species of this genus in the 'Malacological Magazine,' 1838, all of which are northern forms.

1. MARGARITA ELEGANTISSIMA. *Bean*, MS. Tab. XV, fig. 1.

M. Testá orbiculari, conoideá, tenui; apice acuto; anfractibus quinque, supernè convexis, infernè planulatis; longitudinaliter obliquè costulatis; interstitiis tenuissimè striatis; umbilico magno, bicarinato.

Shell orbiculato-conoidal, thin, and fragile, with a slightly elevated spire and acute apex; volutions five, convex above, rather flat beneath, outer edge obtusely angular; longitudinally costated, with elevated, reflected, or imbricated costæ, finely striated transversely; umbilicus large, bicarinated.

Axis, ¼ of an inch.

Locality. Mam. Crag, Bridlington.

The cabinet of Mr. Bean, of Scarborough, has furnished me with the only specimen of this species that I have seen. It appears to be distinct from any with which I am acquainted, and I have, in accordance with that gentleman's opinion, considered it a new species. The characters which most distinguish it are the prominence of the

---

* Etym. *Margarita*, a pearl; from the nacreous character of these shells.

margin surrounding the umbilicus, and its elegantly imbricated costæ, or reflected lines of growth. The spiral striæ are most distinct at the outer part of the base of the shell, but traces of them may be detected on the upper part of the volution between the costæ. The aperture of the shell is hidden in the matrix.

2. MARGARITA (?) MACULATA. *S. Wood.* Tab. XV, fig. 3.

    SOLARIELLA MACULATA. *S. Wood.* Cat. 1842, p. 531, pl. 5. fig. 7, 10.
    TURBO MONILIFERUS. *Nyst.* Rech. Coq. foss. d'Anvers 1835, p. 27.
    SOLARIUM TURBINOIDES. *Nyst.* Coq. foss. de Belg. p. 370, pl. 26, fig. 7, 1844.
    —   *Lyell.* In Quart. Journ. Geol. Soc. p. 555, 1845.

M. Testá trochoideá, vel solariformi, tenui et fragili; spirá depressá; anfractibus quinque, subcylindraceis, crenulatis; cingulis tribus vel quatuor, ornatis; suturis profundis canaliculatis; aperturá rotundatá; umbilico magno, granulato.

Shell depressedly conical, subtrochiform, or rather solariform; spire slightly elevated; apex rather obtuse; volutions four or five, subcylindrical, with three, sometimes four, rugose or crenulated bands or ridges, the upper one of which is the most prominent; base rough, striated or ridged, with a large open umbilicus, and a carinated and granulated edge; shell nacreous.

Axis, $\frac{1}{4}$ of an inch nearly; *diameter,* $\frac{3}{8}$

Locality. Cor. Crag, Sutton.

This shell is given as a Miocene fossil of America by Mr. Lyell. Small specimens and fragments are by no means rare from the Crag; but a shell of the size figured is very difficult to obtain in a perfect condition, from its extreme fragility. It appears, from the form of the shell, to be intermediate between Solarium and Trochus, probably nearly related to the latter; and closely resembling Margarita. It has three large and obtuse ridges or carinæ upon the upper and outer half of the volution, though only two are visible upon the spire. The upper one is largest, and may be said to be divided into three, which are crenulated or granulated by elevated obtuse lines of growth, and there are upon the base about half a dozen ridges or elevated concentric lines; the largest, surrounding the umbilicus, is roughly crenulated, like a Solarium, and the volutions are visible beneath up to the apex. In well-preserved specimens the shell is covered with very fine spiral striæ. A deep and broad canal is formed near the suture by the elevation of the large upper keel, and there are the remains of coloured blotches, from which character its name was given.

The Belgian shell, M. Nyst informs me by letter, is the same as the British fossil; though the figure in the work of that gentleman is much more elevated than any of my specimens.

3. MARGARITA TROCHOIDEA. *S. Wood.* Tab. XV, fig. 2.
MARGARITA TROCHOIDEA. *S. Wood.* Catalogue 1842.

M. Testá orbiculato-trochideá, depressá, tenui, fragili; apice obtusiusculo, anfractibus quatuor, convexis, supernè depressis apud suturas, ad basim obtusè angulatis, subtus rotundatis; transversim tenuissimè striatis; aperturá suborbiculari; umbilico magno profundo.

Shell depressedly conical, trochiform, thin, and fragile; volutions four, convex above, slightly flattened near the suture, convex or rounded beneath, with an obtuse angle at the outer edge; covered with fine transverse or spiral striæ; aperture subcircular, impressed by the body whorl with a sharp expanded and curved outer lip; umbilicus large and deep.

Axis, $\frac{1}{4}$ of an inch; *diameter* $\frac{3}{8}$.

Locality. Cor. Crag, Sutton.

Fragments and small specimens of this species are abundant. The diameter of the base of the specimen figured does not exceed three eighths of an inch, although fragments indicate at least half an inch. The volutions are convex both above and beneath the subcarinated edge, and the umbilicus is visible to the spire, characters that distinguish this species from any other with which I am acquainted.

ADEORBIS,* *S. Wood.* 1842.
HELIX (spec.) *Mont.*
TROCHUS (spec.) *Brown, Nyst.*
CINGULA (spec.) *Flem.*
NATICA (spec.) *Phil.*
VALVATA (spec.) *Phil.*
DELPHINULA (spec.) *Bast.*
MARGARITA (spec.) *S. Wood.*

Gen. Char. Shell generally small, suborbicular, depressed, with a few nearly discoidal and rapidly increasing volutions, umbilicus large and deep; peritreme entire, and nearly continuous, slightly interrupted by the previous volution, deeply sinuated on the inner side, and having a minute or incipient sinus at the upper part of the aperture, near the junction of the body whorl.

As there was not any genus established at the time I compiled my Catalogue in which I could place some small shells from the Crag, possessing the above characters, the above name was proposed for their reception. Conceiving that an animal with

* Etym. *Adeo,* to approach, *orbis,* a circle.

such a form of aperture to its shell would, when known, be found to have characters sufficiently distinct to entitle it to generic distinction.

It was considered as differing from Skenea, in the peculiarly sinuated form of its aperture.

1. ADEORBIS STRIATUS. *S. Wood.* Tab. XV, fig. 7.

ADEORBIS STRIATUS. *S. Wood.* In An. and Mag. of Nat. Hist. 1842, vol. ix, p. 530, pl. 5, fig. 4, 6.
VALVATA STRIATA. *Phil.* En. Moll. Sic. vol. i, p. 147, t. 9, fig. 3, 1836.

A. Testá depressá, minutá, striatá, subdiscoideá; anfractibus tribus, convexis, vel subcylindraceis; spirá vix elevatá, umbilico magno; aperturá patente; veritremá supernè expansá, infernè sinuatá.

Shell depressed, small, and striated; volutions three or four, nearly horizontal; rounded or subcylindrical flattened above; spire scarcely elevated, with a deep distinct suture; umbilicus large and wide; aperture gaping, peritreme projecting on the upper part, deeply sinuated underneath.

Diameter, $\frac{1}{7}$ of an inch; *altitude,* $\frac{1}{2}$ the diameter.

Locality. Cor. Crag, Gedgrave and Sutton.

This is at present a rare shell. I have not met with more than four or five specimens. The volutions are slightly depressed above, and the upper part of the lip is as nearly as possible at the centre of the preceding volution; the shell is regularly covered with about 12—15 broad and rounded striæ; the umbilicus is large, with the volutions visible beneath, up to the apex, and the shell wholly destitute of carinæ. This is probably the same species as M. Philippi's shell, which he considered as a Valvata, in consequence of its association with *Cyrena Gemmallarii*. This Cyrena is also found mixed with marine shells in the Coralline Crag, at Gedgrave. I believe our shell, *A. striatus*, to have no connexion with Valvata, but that all the species I have included in this genus are marine animals.

2. ADEORBIS SUPRA-NITIDUS. *S. Wood.* Tab. XV, fig. 5, *a—b.*

ADEORBIS SUPRA-NITIDUS. *S. Wood.* Catalogue 1842.

A. Testá depressá, minutá, supra-nitidá, lævigatá, politá, carinatá, aliquando tricarinatá; anfractibus 3—4, convexis aut cylindraceis; propè suturam subcanaliculatis; umbilico magno, aperto, et rugosè striato.

Shell depressed, small, smooth, glossy, and naked above, with from one to three sharp carinæ, the upper one small, often wanting; volutions 3—4, with a depression or subcanal near the suture; umbilicus large, open, coarsely striated within; peritreme sharp, slightly interrupted by the body whorl.

Diameter, $\frac{1}{8}$ of an inch; *altitude,* $\frac{1}{2}$ the diameter.

Locality. Cor. Crag, Sutton.

A very abundant shell. It differs from the preceding, *A. striatus*, in being quite free from striæ on the upper part, and in having three keels at unequal distances upon the volution, of which the lower one is the most prominent and permanent, and the upper one, even the upper two, are sometimes evanescent. A small incipient sinus is visible at the upper part of the aperture, which gives a depression round the volution near, but not at, the suture, which character I do not observe in the other species. This may possibly be the *Skenea divisa*, Flem. (Brit. An. p. 314), *Turbo divisa*, Adams; but these authors make no mention of the keel-like ridges that surround the shell.

Three or four coarse striæ or ridges cover the volution at the umbilicus, and the lines of growth are elevated at that part, and are corrugated. The most abundant variety is that which is entirely smooth above, with only one keel upon the base or outer edge of the volution.

3. ADEORBIS TRICARINATUS. *S. Wood*. Tab. XV, fig. 6.
ADEORBIS TRICARINATUS. *S. Wood*. Catalogue 1842.

A. Testá minutá, depressá, subdiscoideá; anfractibus subcylindraceis; cingulis tribus acutis, subequidistantibus, instructis; interstitiis spiraliter striatis; supernè et infernè planiusculis, umbilico magno.

Shell small, depressed, subdiscoidal; with three convex or rather subcylindrical volutions, furnished with three sharp subequidistant and elevated carinæ spirally striated between them; shell rather flat above and beneath, with a large open umbilicus, slightly corrugated by lines of growth.

Diameter, $\frac{1}{8}$ of an inch; *altitude*, $\frac{1}{2}$ its diameter.

Locality. Cor. Crag, Sutton.

I have about half a dozen specimens in my cabinet with the above characters, in which they appear to differ from the two preceding species sufficiently to be considered specifically distinct.

A. supra-nitida possesses the keels, but less prominently, and less equally distributed, and it is perfectly smooth except in the umbilicus. *A. striatus* possesses the striæ, but has no vestige of carinæ; and this shell has not the coarse spiral striæ of the umbilicus. It is possible that they may be only varied forms of one species, but the numerous specimens in my cabinet appear to preserve the above distinctions, and they must for the present, therefore, remain separate.

4. ADEORBIS SUBCARINATUS. *Mont.* Tab. XV, fig. 8.

 HELIX SUBCARINATUS. *Mont.* Test. Brit. p. 438, t. 7, fig. 9, 1803.
 TROCHUS SUBCARINATUS. *Brown.* Conch. Illust. p. 51, fig. 16-17, 1827.
  — RUGOSUS. *Brown.* Wern. Mem. ii, 520, t. 24, fig. 5, (*Flem.*)
 CINGULA SUBCARINATA. *Flem.* Brit. An. p. 305, 1828.
 DELPHINULA TRIGONOSTOMA. *Bast.* Bord. foss. pl. 4, fig. 10, 1825.
 ADEORBIS SUBCARINATUS. *S. Wood.* Catalogue 1842.
 TROCHUS TRIGONOSTOMUS. *Nyst.* Coq. foss. de Belg. p. 385, pl. 35, fig. 23, 1844.
 NATICA SUBCARINATA. *Phil.* En. Moll. Sic. vol. ii, p. 141, t. 24, fig. 13, 1844.

A. Testá pusillá, orbiculato-depressá, spirá brevi ; anfractibus tribus ; subtus planatis ; longitudinaliter lineatis, et transversim cingulatis ; cingulis elevatis ; aperturá magná subtetragoná ; umbilico lato.

Shell small, thick, and strong, with three whorls rapidly enlarging, flattened beneath, covered with four elevated ridges, one of which is in the centre of the base ; obliquely crenulated between the ridges ; aperture large, subquadrangular ; outer lip curved and expanded ; umbilicus large and deep.

Diameter, $\frac{1}{10}$ of an inch.

Locality. Cor. Crag, Sutton.

    Red Crag, Sutton.   Recent, Britain.

This shell is rare in my cabinet. I have no doubt of its identity with the recent British species, and the figures above referred to are presumed to be the same. Our shell has four distinct ridges—three on the upper part of the whorl, at unequal distances, and one upon the centre of the base, with a large and patulous mouth of a somewhat quadrate form, deeply sinuous on the inner side, with an incipient canal both at the upper and lower angles of the mouth ; the upper one raising the shell slightly at the suture, which gives a depression on the volution.

5. ADEORBIS PULCHRALIS. *S. Wood.* Tab. XV, fig. 4.

 MARGARITA HELICINA. *S. Wood.* Catalogue 1842.

A. Testá minutá, tenui, pellucidá (?), depressá, suborbiculari, heliciformi ; anfractibus tribus, supernè convexis, subtus planiusculis, transversim tenuissimè striatis ; aperturá patente, obtusè angulato ; umbilico magno, profundo ; peritremá acutá, supernè incurvá prominente, intus et infra valdè sinuatá.

Shell small, thin, and probably pellucid in the recent state ; depressed and suborbicular, heliciform, very finely striated along the volutions ; volutions three, rapidly increasing ; convex above and rather flat beneath, obtuscly angulated on the outer edge ; aperture large and wide, somewhat angular, with a large and deep umbilicus ; peritreme sharp, slightly interrupted by the volution, curved, and projecting on the upper side, and deeply sinuated on the inner margin.

Diameter, $\frac{1}{7}$ of an inch ; *altitude,* $\frac{1}{2}$ the diameter.

Locality. Cor. Crag, Sutton.

This elegant little shell is not particularly rare. My first specimens were more or less injured at the aperture, and the possession of some, obtained recently, have the peritreme perfectly entire, showing a deeply sinuated form of aperture, like that of *A. subcarinatus.* I have, therefore, placed it among its probable relations.

The apex is somewhat obtuse, and the suture deep, with an incipient canal at the junction of the outer lip with the body whorl, where it joins, as nearly as possible, in the centre of the volution. The striæ are rather coarse, and more distant upon the base of the shell than they are upon the upper part, and the lip is slightly sinuated at the outer or lower angle of the aperture.

NATICA, *Adanson,* 1757.

RUMA (spec.) *Chem.* (fide *Gray.*
UBER. *Humph.* 1797.
LUNATUS. Id.
NATICUS. *Montf.* 1810.
POLINICES. Id.
MAMMILLA. *Schum.* 1817.
NACCA. *Risso.* 1826.
NEVERITA. Id.
AMPULLARIA. *Flem.* 1828.
GLOBULUS. *J. Sow.* 1834.
BULBUS. *Brown.* 1838.
GLOBULARIA. *Swains.* 1840.
MAMMILLARIA. Id.
PACHYLABRA. Id.
NATICINA. *Guild.* 1840.
NATICELLA. Id.
CERNINA. *Gray,* 1840.
AMAURA. *Möller,* 1842.
NATICOPSIS. *M'Coy.*
DESHAYESIA. *Raulin,* 1844.

Gen. Char. Shell generally thick, strong, smooth, and glossy, occasionally covered with fine striæ, of an ovate, globulous, or subspheroidal form, with a short or slightly elevated spire; aperture oval, or semilunate; outer lip plain and simple; inner lip sometimes depositing a callosity, the callus modifying the form and size of the umbilicus; operculum corneous or calcareous.

This is at present by no means a well-defined genus; several species have been separated, in consequence of an extended form of the inner lip, by which the umbilicus is closed, while some naturalists have considered the distinction of a corneous from a calcareous operculum sufficient to justify generic separation. The author of 'Malacologia Monensis' (1838, p. 62), reports that the animals of two calcareous operculated Naticæ, obtained at Algiers, exactly agreed in generic character with the British corneous operculated species; and M. Philippi examined the animals of species possessing a calcareous operculum, and also those with a corneous one, and says (En. Moll. Sic. vol. i, p. 161) that they present a similar structure. The Crag formations contain species that possessed a calcareous operculum, as well as those in which the operculum was, no doubt, of a corneous texture, and they are here considered as constituting only one genus. This genus is placed among the circular mouthed, or what are generally considered as phytophagous animals, although some of the species are not only carnivorous, but exceedingly voracious. Some species are found in deep water, while others bury themselves in sand, above low-water mark. It may be considered as a marine genus, though specimens are occasionally found fossil, in estuary deposits.

SECT. *a*. OPERCULUM CORNEOUS.

1. NATICA CATENOIDES. *S. Wood.* Tab. XVI, fig. 10, *a, b, c.*

NATICA CATENOIDES. *S. Wood.* Catal. in An. and Mag. Nat. Hist. 1842, p. 529.

— GLAUCINOIDES. *J. Sow.* Min. Conch. t. 479, fig. 4, *bene.*

— SOWERBYI (?). *Nyst.* Coq. foss. de Belg. pl. 37, fig. 31, 1844.

N. Testá ovato-globosá, ventricosá, lævi, politá; anfractibus 5—6, tumidis, depressis; spirá productiusculá; aperturá ellipticá, obliquá; umbilico mediocri aperto.

Shell globose, or ventricosely ovate, smooth, and glossy; spire slightly produced; volutions about five, depressed, a little flattened at the upper part; aperture elliptical, with a moderate umbilicus, open and deep; lower part of the left lip slightly reflected and callous.

Longest diameter, 2 inches.

Locality. Red Crag, Sutton and Walton.

This species is by no means rare. I have never seen a specimen that has not in some degree been altered by the removal of its glossy outer coating, under which there is a striated surface, and the specimen figured is in that condition; but portions of that vitreous covering may occasionally be seen, and the shell in its original state was perfectly smooth and polished. The spire is very slightly elevated, with scarcely any depression at the suture, but the upper part of the volution is a little flattened. I have not any specimens that show such an elevated spire and depressed suture as is represented in the Belgian fossil of what I presume to be this species, except those which are much decorticated at the suture. Our shell is more in form like the *N. Alderi*, and might possibly be a large variety of that species. I think it is specifically distinct from *N. catena*, which has a more elevated spire. Among all my specimens there is only one which has the umbilicus perfect, and in it the shell is rather roughly striated, as it is also in *N. catena*, where there is a slight callosity at the lower part of the left lip.

The young of this species so much resembles *N. Alderi*, that it is extremely difficult to distinguish them; and it is possible that species may have been an inhabitant of the Crag Sea, but my cabinet does not contain a good or decided specimen, and this may perhaps be considered as its representative.   Fig. 10 c shows the coarse striæ in the umbilicus.

2. NATICA CATENA. *Da Costa.* Tab. XVI, fig. 8, *a—b.*

COCHLEA CATENA.   *Da Costa.* Hist. Nat. Test. Brit. p. 83, t. 5, fig. 7, 1778.
NERITA GLAUCINA.   *Mont.* Test. Brit. p. 469, 1803.
NATICA BRITANNICA.   *Leach.* MS. 1818.
—   MONILIFERA.   *Lam.* Hist. des An. sans Vert. 2d edit. tom. viii, p. 638, 1838.
—   NICOLII.   *Forbes.* Malac. Mon. p. 30, 1838.
—   GLAUCINOIDES (?).   *Nyst.* Coq. foss. de Belg. p. 442, pl. 37, fig. 32, 1844.
—   CATENA.   *S. Wood.* Catalogue 1842.

N. Testá ventricoso-globosá, lævigatá, politá; spirá prominulá; anfractibus 5—6, convexis, supernè depressiusculis; aperturá ovatá, subrectá; umbilico aperto.

Shell ovato-globose, smooth, and glossy, with a short but distinctly produced spire; whorls 5—6, convex, somewhat flattened above, with a deep suture; aperture ovate; umbilicus open and deep; left lip slightly spreading, and rather callous below.

Longest diameter, 1¼ inch.

Locality.   Red Crag, Sutton.

Mam. Crag, Bridlington.   Recent, British Seas.

This shell is not so abundant as the preceding; but I have at least a dozen specimens in good preservation, and they correspond so precisely with the recent British species, that there is little doubt of the identity.   This differs from the preceding species in having the volutions more convex and distinct, with a more elevated spire, and it is not so large.   These shells are also somewhat altered by decortication, more especially at the suture; but some specimens have portions which sufficiently show the original form, and with a diameter of only half that of the preceding species, they have as many volutions.

"This species is found buried in sands at very low tides in the Frith of Forth." (Forbes's Mal. Monensis, p. 30.)

3. NATICA GUILLEMINI (?). *Payraudeau.* Tab. XVI, fig. 1, *a—b.*

NATICA GUILLEMINI.   *Payr.* Cat. des Moll. de l'Ile de Corse, 1826, p. 119, pl. 5, fig. 25-26.
—   *Desh.* 2d edit. *Lam.* vol. viii, p. 648, 1838.
—   *Phil.* En. Moll. Sic. vol. i, p. 162, 1836.
—   -   -   -   vol. ii, p. 140, 1844.

N. Testá ventricoso-ovatá, glabrá; spirá prominulá, acutiusculá; anfractibus quinque convexis, subdepressis; umbilico parvo, subnudo; labio adnato, calloso.

Shell ventricosely-ovate and smooth, with a slightly prominent spire, and rather acute apex; volutions five, depressedly rounded; and a small umbilicus, partly hidden by the lower portion of the callous and reflected left lip.

Longest diameter, ¾ of an inch.

Locality. Red Crag, Sutton.   Recent, Mediterranean.

A few specimens of this shell in my cabinet appear to resemble the recent Mediterranean species in the peculiar form of the reflected and thickened portion of the lower part of the left lip which extends partly over what would be a small umbilicus, but the specimens are much altered and decorticated, thereby causing a greater opening at that part of the shell. I have considered this as different from the young of *N. catenoides* on account of a greater number of volutions in shells of the same size; a specimen of this species having five volutions, with a diameter scarcely exceeding a third of that of *N. catenoides*, which has never more than six. It must, however, for the present, be considered doubtful.

4. NATICA PROXIMA. *S. Wood.* Tab. XVI, fig. 4, *a—b.*
NATICA PROXIMA. *S. Wood.* Catalogue 1842.

N. Testá subglobosá, obliquá, lævigatá, politá; anfractibus quinque subdepressis; spirá breviusculá; aperturá semilunari; umbilico lato, labio calloso.

Shell subglobose, smooth, and glossy; volutions about five, oblique, and rather depressed, spire short; aperture hemispherical, with an open and deep umbilicus, modified by the callosity of the left lip.

Longest diameter, 1 inch.

Locality. Cor. Crag, Ramsholt.

My cabinet contains but a few specimens of what I have presumed to be a distinct species. The umbilicus is large and deep, somewhat reduced above by the callosity of the lower portion of the left lip, and there is a peculiar flatness upon the columella, unlike the line produced by the pressure of a testaceous operculum. Some faint lines of striæ are visible when the outer covering is removed, particularly near the suture. It differs from *N. varians* in having a much less elevated spire, a less depressed suture, and a rather larger callosity. It somewhat resembles *N. hemiclausa*, but that species is rather more depressed in the volutions, and has its umbilicus entirely closed.

5. NATICA VARIANS. *Dujard.* Tab. XVI, fig. 6, *a—b.*
NATICA VARIANS. *Dujard.* Mém. de la Soc. Géol. de France, tom. xi, pt. 2, p. 281, pl. 19, fig. 6, 1837.
— HEMICLAUSA (?). *Nyst.* Coq. foss. de Belg. p. 446, pl. 38, fig. 15, 1844.
— ELEVATA. *S. Wood.* Catalogue 1842.

N. Testá ovato-conicá, crassá, lævigatá, politá, spirá elevatá; anfractibus sex obliquis; aperturá seminulatá, labio calloso; umbilico magno aperto, profundo.

Shell ovato-conical, thick, strong, smooth, and glossy, with an elevated spire, and about six oblique or sloping volutions; umbilicus rather large, open and deep; left lip callous.

Longest diameter, $1\frac{3}{8}$ inch.

Locality.   Cor. Crag, Ramsholt.

　　　　　Red Crag, Sutton.

This is not a very abundant shell in my cabinet from either formation.   It appears to be quite distinct from *N. hemiclausa,* and agrees in most of its characters with *N. varians,* from Touraine.   In this shell the umbilicus is tolerably large and deep, with a very slight ridge at the lower part, which is perceptible in some of the specimens from Touraine, lent me for comparison by Mr. Lyell.   The shell figured by M. Nyst is probably a variety of this, but the volutions of my specimens are more depressed ; the left lip is callous, particularly at the upper angle of the aperture, and it never covers the umbilicus.   A slight flattening is visible on the left side, but apparently not deep enough for the pressure of a calcareous operculum, nor is there any mark within the outer lip.   This shell is striated beneath the outer coating.   It somewhat resembles in form *N. plumbea,* Lamarck, but it is a thicker shell, more elongated, and has a more distinct and elevated spire.

6. NATICA HEMICLAUSA.   *J. Sow.* Tab. XIV, fig. 5, *a—b.*

　　　　NATICA HEMICLAUSA.   *J. Sow.* Min. Conch. t. 479, fig. 2, 1824.

　　　　　　—　　　　　　 *S. Wood.* Catalogue 1842.

　　　　　　—　　　　　　 *Morris.* Catalogue of Brit. Fossils, p. 153, 1843.

N.   Testá ovato-conicá, obliquá, lævigatá, politá ; spirá conoideá ; anfractibus sex depressis ; aperturá ovatá ; labio calloso ; umbilico tecto.

Shell ovato-conical, oblique, smooth, and polished ; spire conical ; whorls six, depressed ; aperture ovate, with a callous left lip closing the umbilicus.

Long. diameter, 1 inch.

Locality. Red Crag, Walton Naze and Sutton.

This is also an abundant shell.   The umbilicus is quite covered when the animal has reached maturity.   The specimen figured in ' Min. Conch.' was a young individual. It much resembles *N. mammilla,* Lamarck, but its left lip is less callous, especially at the upper part, and it has a rather more prominent spire ; it was smooth and glossy, and, like *N. mammilla,* has the volutions equally depressed.   It also shows a deepened suture when the outer covering has been removed, a condition in which the Crag Naticæ are often found.   M. Philippi considered this, in his first vol. p. 162, to be the same species as *N. Guillemini,* Payr., and, in vol. ii, p. 140, he has identified it (doubtfully) with *N. macilenta* (t. 24, f. 14).   I think, however, the Crag shell sufficiently distinct from either.   Specimens are occasionally more elongate than the one represented.

7. Natica cirriformis. *J. Sow.* Tab. XVI, fig. 7, *a—b.*

Natica cirriformis.   *J. Sow.* Min. Conch. t. 479, fig. 1.
—   *S. Wood.* Catalogue 1842.
—   *Morris.* Catalogue of Brit. Fossils, p. 152, 1843.
—   *Nyst.* Coq. foss. de Belg. p. 444, pl. 39, fig. 1, 1844.

N. Testá globosá, crassá, lævigatá; anfractibus sex, convexis; spirá depressá; aperturá semilunatá; labio infernè calloso; umbilico lato, profundo.

Shell globose, thick, smooth; volutions convex, tumid; spire slightly elevated; umbilicus wide and deep, with a broad sinus in the left lip, callous at the lower part; aperture semilunate.

Longest diameter, 1¼ inch.

Locality. Cor. Crag, Ramsholt.

I have not seen this species from the Red Crag; and specimens of it from the Coralline beds are generally in a decorticated state. Its distinguishing character is the callosity at the lower part of a large umbilicus, the upper side of which has a distinct ridge when the shell is well preserved, the left lip being, as it were, separated in the middle by a deep and broad canal; unlike *N. multipunctata,* the ridge is on the lower part of the umbilicus, and not in the middle; the volutions expand, and are somewhat compressed above and below, forming a large and wide umbilicus.

8. Natica helicoides. *Johnston.* Tab. XVI, fig. 3, *a—b.*

Natica helicoides. *Johnston.* Hist. of the Berwickshire Nat. Hist. Club, 1834.
—   —   *Lyell.* In Mag. Nat. Hist. 1839.
— canaliculata. *Gould.* Report on the Inv. of Massachusetts, p. 235, fig. 161, 1841.
— helicoides. *S. Wood.* Catalogue 1842.
—   —   *Morris.* Catalogue of Brit. Fossils, p. 153, 1843.

N. Testá ovatá, crassá, lævigatá (?); spirá elevatá; anfractibus quatuor vel quinque convexis; suturis profundis; aperturá ovatá; umbilico tecto.

Shell ovate, thick, and smooth (?); spire elevated; volutions four or five, convex, with a deep or canaliculated suture; aperture ovate, scarcely oblique, and without an umbilicus.

Axis, 1 inch nearly.

Locality. Red Crag, Sutton.
Mam. Crag, Bramerton and Bridlington.
Recent, North Seas and Coast of Massachusetts.

This shell is rare in the Red Crag, though I believe it is not so in the Mammaliferous formation. It was given as an extinct species in the list of Mammaliferous Crag Shells, published in the 'Mag. Nat. Hist. 1839', the joint production of Mr. Lyell, Mr. G. Sowerby and myself, as we were not then aware of its existence as a recent shell. We have since been able to compare the recent and fossil shells, and have no

19

doubt of the identity.   The original specimen dredged up to the northward is covered with fine striæ, but the fossils are quite smooth.   An imperfect specimen is among the shells sent by Mr. Leckenby, found by that gentleman at Bridlington.

9. NATICA OCCLUSA. *S. Wood.* Tab. XII, fig. 4, *a—b.*

N. Testá globosá, crassá, imperforatá, lævigatá; spirá elevatá; anfractibus quinque, convexis; suturis profundis; aperturá ovatá; umbilico callo, complanato, prorsus obtecto.

Shell globose, imperforate, thick, strong, smooth, and glossy, with five or six convex volutions, and a deep distinct suture; spire elevated; apex somewhat acute; aperture ovate; umbilicus covered by the callosity of the left lip.

Axis, 1 inch.

Locality.   Mam. Crag, Bridlington.

One specimen of this shell, from the cabinet of Mr. Leckenby, is all that I have seen; and being unable to identify it with any species known to me, I have ventured to give it a name.   Its most distinguishing character is the callosity of the left lip, which entirely covers the umbilicus much in the manner of that in *N. clausa*, from which, however, it appears to differ in its much more elevated spire and less tumid volutions, and it also wants the peculiar flatness near the suture, which exists in that species.   There is a slight flattening upon the lower part of the columella, but not the distinct impression on either side of the aperture that is generally made by the pressure of a calcareous operculum; I therefore presume this to have had a corneous one.   It is faintly but distinctly covered with fine spiral striæ upon its outer coating.

10. NATICA GRŒNLANDICA. *Beck.* Tab. XII, fig. 5, *a—b.*

    NATICA GRŒNLANDICA.   *Beck* (in *Möller*). Ind. Moll. Grœnland. p. 7, 1842.

    —   —   *Thorpe.* Brit. Mar. Moll. fig. 55, 1844.

    — LIVIDA. *Bean* (in *Thorpe*). p. 265.

    — GRŒNLANDICA. *Forbes.* Rep. upon the Geol. Rel. of the exist. Faun. and Flor. of
                 the Brit. Isles.

N. Testá globosá, læviusculá, solidiori, perforata; anfractibus quatuor, convexis; spirá mediocriter elevatá; apice obtuso, plerumque corroso; peristomate subcontinuo; umbilico angusto, profundo; operculo corneo, tenui.

Shell globular and rather glossy, strong, and perforated with four convex volutions, and a slightly elevated spire, finely striated transversely; umbilicus small, deep, and open; left lip slightly spreading; aperture ovate; operculum thin and corneous.

Axis, ½ an inch.

Locality. Mam. Crag, Bridlington.

                           Recent, Coast of Scarborough and North Seas.

The beautiful specimen from which the figure above referred to was taken is from the Museum of the Philosophical Institution, at York, obligingly sent me by Mr. Charlesworth, for the purpose of being described. I think that there is little doubt of this being identical with the recent Greenland species. Fine striæ cover the exterior, which are most visible near the suture and upon the base; and the apex, though not much eroded, has lost its outer coating.

Sect. β. OPERCULUM CALCAREOUS.

11. **Natica clausa**. *Broderip* and *Sowerby*. Tab. XVI, fig. 2, *a—b*.

Natica clausa. *Brod.* and *Sow.* Zool. Journ. No. 15, p. 372, 1829.
— *Gray.* Zool. of Beechey's Voy. pl. 37, fig. 6, 34, f. 3.
— *Smith.* Mem. Wern. Soc. vol. viii, p. 55, pl. 1, fig. 16, 1838.
— *Gould.* Invert. of Massachus. p. 238, fig. 167, 1841.
— *S. Wood.* Catalogue 1842.
Natica consolidata. *Couthouy.* Bost. Journ. Nat. Hist. xi, p. 89, pl. 3, fig. 14.
— septentrionalis. *Möller.* Ind. Moll. Grœn. p. 7, 1842.

N. Testá globosá, imperforatá, lævigatá ; anfractibus quinque ventricosioribus ; propè suturam planulatis ; spirá parum elatá ; aperturá ovatá ; umbilico callo semiorbiculari, complanato, prorsus obtecto.

Shell globose, imperforate, smooth; spire slightly elevated, with five convex and rather tumid volutions, somewhat flattened on the upper part; aperture ovate, left lip with a semiorbicular and flattened callosity, completely covering the umbilicus.

Axis, 1 inch nearly.

Locality. Red Crag, Sutton.

Mam. Crag, Bridlington.

Recent, North of Europe, and Seas of Boreal America.

This shell is rare as a Red Crag fossil. My cabinet contains but two specimens, which, however, correspond precisely with the figure and description above referred to, and I have no doubt of the identity. My specimens show a few traces of fine striæ, where the outer covering has been removed, and there is the impression of its calcareous lid on both sides of the aperture. It is very common in some of the Swedish beds, as mentioned by Mr. Lyell, and is quoted by Professor E. Forbes, in his ' Report on the Existing Fauna and Flora of the British Isles,' as found at Bridlington, though I have not seen a specimen from that locality.

12. NATICA MULTIPUNCTATA. *S. Wood.* Tab. XVI, fig. 9, *a—f.*

> NATICA PATULA.   *J. Sow.* Min. Conch. t. 373.
> — — *Nyst.* Rech. Coq. foss. d'Anvers, 1835, p. 25, No. 12.
> — — *Morris.* Cat. of Brit. Fossils, p. 153, 1843.
> — MULTIPUNCTATA. *S. Wood.* Catalogue 1842.
> — CRASSA (?). *Nyst.* Coq. foss. de Belg. p. 443, pl. 37, fig. 33, 1844.

N. Testá ellipticá, obliquá, depressá, crassá, lævigatá, multipunctatá; spirá produc-tiusculá; anfractibus quatuor, supernè subdepressis, infernè expansis; labio calloso, crasso; umbilico spiraliter jugoso; operculo ad margine unisulcato.

Shell ovate, depressed, oblique, with an elliptical outline; thick, and smooth, covered with numerous spots; spire slightly produced; upper part of volution a little depressed, mouth much expanded below; left lip callous, with a small spiral ridge in the umbilicus. Operculum with a deep spiral sulcus near the outer edge.

Longest diameter, 1½ inch.

Locality.   Cor. Crag. Ramsholt and Gedgrave.

Red Crag, Walton Naze.

A most abundant shell at Walton Naze. It differs (as I have before pointed out in my Catalogue in the An. and Mag. Nat. Hist. 1842, p. 529,) from *N. mille-punctata* in the greater size and peculiar flatness of the callosity at the upper part of the umbilicus, while the spiral ridge is much smaller, varying not only in the different distribution of this calcareous deposition at its large and open umbilicus, but also in the greater quantity of it: the volutions are also more depressed, the shell more regularly ovate, and the mouth larger, with a less elevated and distinct spire. The differences here pointed out appear to me sufficient to constitute a specific character, and I have, in consequence, retained my own name, which was published in 1842, in preference to that by M. Nyst, which was published in 1844. The Italian and Bordeaux fossils in my possession differ also from the Crag shell. The smaller figure, 9, *c*, represents a specimen from the cabinet of Miss Alexander, of Goldrood, near Ipswich, who has obligingly permitted me to have it figured. It contains its operculum in position, and was found at Walton-on-the-Naze. This appendage differs from that of *N. millepunctata* in having a ridge with a broad and deep sulcus near the edge; and it is rather singular that this is the only instance, to my knowledge, of this thick and strong operculum having been found, although the shell may be procured by hundreds. Fig. 9, *d e*, is from the Coralline Crag, and is probably only a variety of this species. It had been considered distinct, and possessed in my cabinet the name of *consors*, but the discovery of an operculum from that formation, corresponding with that found *in situ* from the Red Crag, has given reason to suppose them the same species. The greatest difference is in its having a rather more distinct and elevated spire, with a sharper apex, and a rather less depressed or elongated aperture. The impression of a calcareous lid is left upon the shell most distinctly on both sides of the aperture, and I have presumed fig. 9, *f*, to have been the operculum.

NATICA DEPRESSULA.—Some small specimens in my cabinet, not more than a line in diameter, passed under this name in my Catalogue. They are depressed in the volution with a deep and distinct suture, that does not appear to have been produced by erosion, and differ in that respect from the young of some well-determined species that I have traced downwards to no greater magnitude. They must, however, for the present, be considered of doubtful character.

SIGARETUS, *Adanson*, 1757.
STOMATIA. *Browne*, 1756 (fide *Gray*).
CRYPTOSTOMA. *De Blainv.* 1825.

Gen. Char. Shell auriform, or ear-shaped, suborbicular, oblique, with a depressed spire, and consisting of few and rapidly increasing volutions, striated or reticulated upon the exterior; aperture much dilated, wider than long; peritreme disconnected by the body whorl, with, generally, a small but open umbilicus, and a thin corneous operculum.

The shells of this genus are said to be entirely enveloped by the mantle of the animal, although the exterior is covered with striæ or reticulations. It is distinguished from *Stomatia*, Lamarck, by the texture of the shell, which in this is not nacreous, and from *Marsenia*, to which it approaches nearer in form, by its thicker and more opaque shell, and by its having the inner lip spread over the body whorl, forming an umbilicus. In *Marsenia* the left side is sharp and truncated, visible internally up to the spire.

All the known recent species of this genus are natives of warm climates, and two or three species have been found in the Eocene formations. It is first quoted from the Devonian period.

1. SIGARETUS EXCAVATUS. *S. Wood.* Tab. XV, fig. 11.

S. Testâ minutâ, auriformi, tenui, fragili; spirâ acutâ, prominulâ; anfractibus tribus, depressis, obliquis, supra convexis, subtus planatis, imbricatis; ad marginem obtusè angulatis, transversim undulato-striatis, striis exilissimis; aperturâ patente; umbilico magno, profundo.

Shell small, thin, and fragile, with a slightly prominent spire, and acute apex; composed of three depressed volutions, convex on the upper part, and flat beneath, margin obtusely angulated : spirally striated, with fine undulating striæ ; rough and imbricated beneath, aperture expanded ; umbilicus wide and open.

Greatest diameter, 1 line.

Locality.   Cor. Crag, Sutton.

I have only two small specimens of this delicate and tender shell, which may possibly be the young of some larger species ; it somewhat resembles *Sig. Leachii* (Sowerby's Gen., fig. 3), *Cryptostoma Leachii*, Blain., but it appears to differ in several particulars : the outer margin of the volution is more angulated, forming a flattish base, which is rough and strongly imbricated by reflected lines of growth, and the umbilicus is large, open, and visible up to the apex.

MARSENIA, *Leach*, 1820.

BULLA (spec.)   *Mont.*
LAMELLARIA (spec.)   *Mont.*
CORIOCELLA.   *De Blainville,* 1824.
CHELINOTUS.   *Swains.* 1840.
SIGARETUS *Flem. Phil. Thorpe.*

Gen. Char.   Shell internal, convolute, ear-shaped, thin, delicate, fragile, pellucid, semipapyraceous ; consisting of a few rapidly increasing volutions, and an expanded aperture, with a small depressed spire ; peritreme sharp and thin, confluent with the columella, which is visible internally up to the spire ; shell wholly enveloped by the mantle of the animal.

In my Catalogue the above generic term was employed, in consequence of Dr. Leach having given that name to the well-known British species, *Bulla haliotoidea,* Mont., in his 'Mollusca,' a part of which was printed in 1820.   A species of this genus was described also by Montague (in the Linn. Trans., 1815, vol. ii, p. 186), under the name of Lamellaria, and included with an animal of quite a different form, which he had considered as its type, from which therefore it must be removed.   In 1825 M. de Blainville (in his System of Malacology) proposed a genus under the name of Coriocella ; but in his description of *C. nigra,* from the Isle of France, the animal he considered as the type of his genus, he states that no trace whatever of a shell could be discovered, " sans trace de coquille extérieur ni intérieur" (p. 466).   " Specimens of *Coriocella nigra* in the British Museum, presented by Cuvier, and described by De Blainville, have a distinct shell." (Vide Gray, Zool. Proc., 1847, p. 143.)

As the name Marsenia was given to a determined species, and published with the ulterior intention of characterizing such genus, it is I conceive the one that ought to be adopted.

1. Marsenia tentaculata. *Mont.* Tab. XV, fig. 10.

    Lamellaria tentaculata. *Montague.* In Linn. Trans. vol. ii, p. 186, t. 12, fig. 5-6, 1815.
           —   *Lovén.* Ind. Moll. Scand. p. 15, 1846.
    Marsenia depressa. *S. Wood.* Catal. in An. and Mag. Nat. Hist. 1842, p. 528, t. 5, fig. 8-9.
    Sigaretus tentaculatus. *Thorpe.* Brit. Marine Conchology, p. 155, t. 1, fig. 3, 1844.

N. Testá minutá, ovatá, auriformi, tenui, fragili, pellucidá (?), lævigatá, politá; spirá depressá; anfractu convexiusculo; aperturá patente; labro acuto arcuato; labio tenui reflexo.

Shell small, ovate, or ear-shaped, thin, fragile, and pellucid (?), smooth and glossy, with one depressed or slightly convex volution; spire rather acute, with a large expanded aperture, and curved outer lip; columella, sharp and truncate; inner lip thin and extended, covering the umbilicus.

Greatest diameter, $\frac{1}{8}$ of an inch.

Locality. Cor. Crag, Sutton.   Recent, British Seas.

Two small specimens of this elegant and fragile shell are all that my cabinet contains. They are probably the young state of *Lamellaria tentaculata*, Mont., and appear to resemble the figure of that species given by Mr. Thorpe, above referred to, except that the outer part of the lip of our shell is not quite so much raised. There is a slight depression behind the sharp edge of the columella, which is visible within the shell up to the apex; the inner lip is thin, folded over this sharpened edge, and spread upon the body of the shell to some distance. The contour of our shell is a trapezoidal form, with the angles rounded off.

Velutina,* *Fleming,* 1820.

Helix (spec.) *Linn.*
Bulla (spec.) *Müller.*
Velutina. *Gray,* 1821.
   —   *De Blainv.* 1825.
Galericulum. *Brown,* 1827.

Gen. Char. Shell thin, obliquely convolute, with very few volutions, rapidly enlarging; spire short, slightly elevated, externally smooth, or finely striated, covered, in a recent state, with a thick epidermis; aperture large, ovate, or suborbicular; outer lip sharp, with a continuous peritreme.

Shells of this character were erected by Dr. Fleming into a genus, the particulars of which were given in the 'Edinburgh Encyclopædia,' vol. xiv, p. 626, published, as I am informed, by Dr. Fleming, in 1820. In the following year, 1821, Mr. Gray employed this name generically, and in 1825 M. de Blainville (in his Malacologie,

* Etym. From the specific name of *Bulla velutina* (Müller), the type of the genus.

p. 468,) made use of the same name, apparently without being aware of its having been previously adopted by Dr. Fleming.

Shells of this genus approach nearest in form to those of Sigaretus; but the volutions in the latter genus are more depressed, and the animal partly envelopes the shell. In this genus the shell is said to be wholly external. I am not aware of any species belonging to this genus of an earlier geological date than the Coralline Crag or Miocene period.

1. VELUTINA LÆVIGATA. *Linn.* Tab. XIX, fig. 8, *a—b.*

> HELIX LÆVIGATA. *Linn.* Syst. Nat. p. 1250, 1766.
> BULLA VELUTINA. *Müll.* Zool. Dan. iii, t. 101, fig. 1-4, 1773.
> HELIX LÆVIGATA. *Penn.* Brit. Zool. t. 86, fig. 139, 1776.
> — *Turt.* Linn. vol. iv, p. 540, 1806.
> — Id. Conch. Dict. p. 69, 1819.
> VELUTINA VULGARIS. *Flem.* Edin. Ency. vol. xiv, p. 626, 1820.
> — CAPULOIDEA. *Blainv.* Malac. p. 469, pl. 42, fig. 4, 1824.
> GALERICULUM LÆVIGATUM. *Brown.* Illust. Brit. Conch. pl. 38, fig. 35, 1827.
> VELUTINA LÆVIGATA. *Flem.* Brit. Animals, p. 326, 1828.
> — RUPICOLA. *Conrad.* Journ. Ac. Nat. Sc. vi, 266, pl. 5, fig. 17-18.
> — LÆVIGATA. *Gould.* Invert. of Massach. p. 241, fig. 159, 1841.
> — — *S. Wood.* Catalogue 1842.

V. Testá auriformi, tenui, fragili, ovatá, inflatá, tenuissimè striatá; anfractibus tribus, convexis, tumidis; suturis profundis; aperturá valdè dilatatá; umbilico tecto.

Shell ovate, inflated, thin, and fragile, finely striated transversely; volutions three, with a deep suture, and a large expanded aperture, subcircular; left lip reflected, covering the umbilicus.

Axis, $\frac{1}{2}$ an inch nearly.

Locality.　Mam. Crag, Thorpe.　　　　　　　　　　Recent, British Seas.

This shell, I believe, is rare, perhaps from its extreme fragility. My cabinet contains but one specimen, which differs in some degree from the recent shell, the outer lip being more elevated, the spire depressed, and the exterior smooth; but in the recent specimens the striæ are scarcely visible when the epidermis is removed. Pennant calls this a fresh-water shell, and Dr. Turton says "it inhabits ponds of Europe." It appears to have been associated with estuary shells, though Montague says it is truly marine, and generally obtained by the dredge in deep water. Dr. Gould says it is found on the beach in sea weed, and also in the stomachs of fishes. *Sigaretus similis*, Woodward (Geol. of Norf. t. 3, f. 8), may probably be this species; but from the figure, it is difficult to determine, as both this and the following are found at Bramerton, the locality whence he obtained most of his fossils.

2. VELUTINA UNDATA (?). *Smith.* Tab. XIX, fig. 10, *a—b.*

    VELUTINA UNDATA. *Smith.* Mem. Wern. Soc. vol. viii, p. 54, pl. 1, fig. 15, 1838.

    GALERICULUM OVATUM (?). *Brown.* Illust. Brit. Conch. pl. 38, fig. 27, 1827.

    VELUTINA ZONATA. *Gould.* Inv. of Massach. p. 242, fig. 160, 1841.

    — ELONGATA. *S. Wood.* Catalogue 1842.

V. Testá elongato-ovatá, tenui, fragili ; spirá vix elevatá ; transversim obsoletè striatá ; anfractibus duobus, convexis ; aperturá ovatá ; labio recurvo, planato.

Shell ovate, very thin, and fragile, with a slightly elevated spire ; volutions two, obsoletely striated transversely ; aperture ovate, slightly contracted at the lower part ; left lip reflected and flattened, or rather concave, extending nearly to the base.

Axis, $\frac{3}{8}$ of an inch.

Locality. Mam. Crag, Bramerton.   Recent, Britain (?), Coast of Massachusetts.

This shell, I believe, is also rare. I have only three specimens in my cabinet which appear to agree with the copious description and beautiful figure by Dr. Gould ; so that I have presumed the Crag shell to be the same as the recent species, although I have been unable to obtain a specimen for comparison. As that gentleman states, this is, in all probability, *Galericulum ovatum*, Brown, which has the prior name, and must be restored if the two shells hereafter prove to be the same.

My specimens are thin and fragile, with the surface rather uneven ; the lines of growth are distinct and irregular, with only faint traces of the spiral striæ that once covered it. The flat part of the inner lip appears to have been a little broader than is represented in the American shell, but that is a character which I think is not sufficient to remove it from the species ; and the lower part of the aperture appears somewhat more rounded. My specimens have little more than two volutions, which are very oblique, and the spire is invisible when looking directly at the aperture of the shell.

3. VELUTINA VIRGATA. *S. Wood.* Tab. XIX, fig. 9, *a—b.*

V. Testá auriformi, ovatá, tenui, fragili ; spirá prominulá ; anfractibus 2—3, tumidis, inflatis, transversim striatis ; suturis profundis ; aperturá ovatá, ad basim angustiore ; labio recurvo planato.

Shell ear-shaped, thin, and fragile ; spire slightly elevated, with two or three inflated volutions, covered with spiral striæ ; suture deep ; aperture ovate, narrower at the base ; with a reflected left lip, and flattened columella.

Axis, $\frac{1}{4}$ of an inch.

Locality. Cor. Crag, Sutton.

About half a dozen specimens of this shell in my cabinet appear to differ from any recent species with which I am acquainted, more especially in the form of the aperture.

The external striæ are rather broad but a little elevated, and the spire is slightly prominent, with the upper part of the volution somewhat projecting, giving the outer lip a straighter form than in the other species; the inner lip or columella is somewhat oblique, flattened, and narrower than in the preceding species, with a less regularly ovate aperture; the outer lip is less curved, giving a slightly angular form to the opening.

Some of my specimens are very imperfect; more and better materials are required to determine the species. From its difference of form I have presumed it to be new, though its present name may be considered only as a provisional one.

CAPULUS,* *De Montfort,* 1810.

> PATELLA (spec.) *Linn.*
> PELEOPSIS. *Lam.* 1815.
> AMALTHEA. *Schum.* 1817.
> ACTITA. *Fisch.* (*Hermansen.*)
> PILOPSIS. *Konig.*
> BROCCHIA. *Bronn.*
> ACROCULIA. *Phillips.*
> CYRTHOLITHES. *Vanuxem.*

Gen. Char. Shell obliquely conical, with a broad expanded base, and a slightly recurved apex; aperture large, generally elliptical, sometimes circular; muscular impressions united posteriorly; rounded and enlarged in front, with an opening between them for the head of the animal; in a recent state, covered with a thick epidermis.

This genus is well distinguished from Patella by the position of the apex, which is situated on the posterior part of the shell, and is generally a little recurved, whereas, in Patella, it is on the anterior, and is simply conical; the muscular impressions, or rather impression (for the two are united into one ligulate band), is therefore on the posterior part, with an opening between the obtuse terminations, for the egress of the animal's head, and is joined behind under the apex. Species from the Palæozoic Rocks have been formed into a genus by Phillips, under the name Acroculia, in consequence of an obliquity in its recurved apex, which character alone is insufficient for generic distinction, and from the excessive variableness of some of my Crag fossils, such character may be almost doubted as sufficient for even specific determination.

* Etym. (?) *Capulus,* a handle.

4. CAPULUS UNGARICUS. *Linn.* Tab. XVII, fig. 2, *a—g.*

  var. α. VULGARIS, fig. 2, *d, e,* Red Crag, Walton.
   β, REGULARIS,   2, *a,*   Cor. Crag, Ramsholt.
   γ, UNGUIS,   2, *b, c,* Red Crag, Sutton.
   δ, ELATA,   2, *f,*   Red Crag, Sutton.
 PATELLA UNGARICA. *Linn.* Syst. Nat. p. 1259.
   — HUNGARICA. *Penn.* Brit. Zool. t. 90, fig. 147.
   — UNGUIS. *J. Sow.* Min. Conch. t. 139, fig. 7, 1816.
   — UNGARICA. *Turt.* Conch. Dict. p. 140, fig. 76, 1819.
 PILEOPSIS UNGARICA. *Brown.* Illust. Brit. Conch. pl. 37, fig. 19-20, 1827.
   —   *Lam.* Hist. des An. sans Vert. 2d edit. tom. vii, p. 609, 1836.
   —   *Dujard.* Mém. Soc. Géol. de France, t. 11, pt. 2, p. 274, 1837.
   —   *Nyst.* Coq. foss. de Belg. p. 355, pl. 35, fig. 8, 1844.
 CAPULUS HUNGARICUS. *Flem.* Brit. An. p. 363, 1828.
 CAPULUS UNGARIOUS. *G. Sow.* Genera of Shells.
   —   *S. Wood.* Catalogue 1842.
   —   *Morris.* Catal. of Brit. Foss. p. 141, 1843.
 PILEOPSIS UNGARICA. *Thorpe.* Brit. Mar. Conch. p. 135, 1844.

C. Testá variabili, sæpe orbiculato-conicá, acuminatá; longitudinaliter striatá; vertice hamoso, involuto; aperturá transversim plerumque latiore.

Shell conical and acuminated, with a suborbicular base, generally wider than long, sometimes longer than wide, longitudinally striated, with a hook-shaped or involute vertex.

Transverse diameter of base 2¼ *inches.*

Locality. Cor. Crag, Sutton, Ramsholt, Gedgrave.

   Red Crag, Sutton, Walton, Newbourn, Bawdsey.

               Recent, British Seas.

This is an exceedingly variable species: in one variety the cone is elevated with little or no inflection of the vertex, while, in another, it is upon a level with the base. In var. *unguis,* fig. 2, *b, c,* the aperture is elongato-ovate, with the vertex recurved, and hanging over the posterior margin. In var. *elata,* fig, 2 *f,* the vertex is nearly central, with scarcely any curvature; the more ordinary form has the base with the transverse diameter the wider. In the depressed form which Mr. J. Sowerby has described as *P. unguis,* the shell is slightly oblique, with an obtuse angular edge upon the back. The striæ are distinctly visible in all these varieties. The margin of this species is frequently distorted, with a sinus or indenture, which is generally on the dextral side. The ligulate band of its deeply-impressed muscle mark extends over more than two thirds of the circumference, with large and obtuse terminations; these are equidistant on each side. In the oblique specimens they appear to be less equal, in consequence of one side being more convex or extended than the other.

2. CAPULUS OBLIQUUS. *S. Wood.* Tab. XVII, fig. 1, *a—b*.

CAPULUS OBLIQUUS. *S. Wood.* Catalogue 1842.

C. Testá elongato-ovatá, obliquá, distortá, subdepressá, lævigatá, politá; vertice involuto, laterali; margini postico superposito.

Shell with an elongato-ovate and oblique opening, externally smooth and glossy, one side truncate, with a depressed and involute apex overhanging the posterior margin.

Longest diameter of base, 1½ inch.

Locality. Red Crag, Walton Naze and Newbourne.

Not a very rare shell, though large and rather tender. It appears to differ from the preceding species in its greater obliquity and smooth exterior. My specimens are without a vestige of striæ; but Crag shells are often so much altered in that respect as not to be depended upon for such characters; the right side, however, in this species is always smaller and truncated, while the left is expanded, and more or less rounded. In all my varieties of *C. ungaricus* I have not observed this regularity of form; and I have no specimens by which they can be connected; it is nevertheless a doubtful species. The great persistency in its obliquity, which may be seen in numerous specimens, appears its greatest claim to specific distinction. It has probably lost a portion of its outer coating.

3. CAPULUS MILITARIS. *Mont.* Tab. XVII, fig. 3, *a—f*.

CAPULUS MILITARIS. var. β, partim sinuosus, fig. 3 *a, b*.
PATELLA MILITARIS. *Mont.* Test. Brit. p. 488, t. 13, fig. 11, 1803.
CAPULUS MILITARIS. *Flem.* Brit. An. p. 364, 1828.
— RECURVATUS. *S. Wood.* Catalogue 1842.
PILEOPSIS MILITARIS. *Thorpe.* Marine Conch. p. 135, 1844.
— *Phil.* En. Moll. Sic. vol. ii, p. 92, 1844.

C. Testá obliquè-conicá, striatá, decussatá; vertice hamoso, revoluto, ad dextram incurvato; margini postico impendente; aperturá elongato-ovatá, irregulari.

Shell obliquely conical, striated, and decussated (?), with the vertex much reflected and incurved; volution inclining to the right; apex overhanging the base of the shell, with an elongate and ovate opening, but very irregular.

Altitude, 1 inch; *length of base,* ⅞.

Locality. Cor. Crag, Sutton.

Red Crag, Newbourn, Sutton, and Walton. Recent, British Seas.

This is a very variable species. I have included in it only those specimens which appear to have a small oblique volution, with the vertex always inclining to the right when the shell is upon its base in its natural position, with the anterior margin forwards.

The aperture is generally longer than it is wide, sometimes twice as long ; a few distorted specimens, however, have the aperture transverse. In its young state, the volution of the vertex is on a level with the margin, projecting beyond it ; and this character is sometimes preserved throughout its existence, though, in some large specimens, the vertex is elevated considerably above the level of the base. It is covered with coarse and prominent striæ, amounting almost to ribs, with deep sulci between them, and occasionally decussating lines of growth, which are, however, very irregular, and not always visible. The muscular band appears like that in *C. ungaricus*, but in my specimens it is not always distinct.

Small and imperfect specimens are by no means rare in the Coralline Crag. In *Capulus ungaricus* the vertex is generally regular and straight, though occasionally it may be observed to incline on one side. This species appears to have a longer cone, with the vertex always on the right side. The sinuses in variety β appear as if they had been produced by the inequalities or rays of a pecten, upon which the animal might have adhered. In my specimens (only three in number) they are upon the left side of the shell, and regular, and if produced in this manner, the animal must have had the power to elevate its right side so as not to be similarly affected, as that side is free from these inequalities ; the striæ upon the exterior take a longitudinal or rather a radiating direction from the vertex, and run obliquely across these sinuses, from which it is presumed they are accidentally produced.

4. CAPULUS FALLAX. *S. Wood*. Tab. XVII, fig. 4, *a—b*.
CAPULUS FALLAX. *S. Wood*. Catalogue, 1842.

C. Testá suborbiculatá, depressá, lævigatá, tenui, fragili ; vertice obliquissimè ad dextram revoluto, spiraliter intorto ; margini postico impendente ; basi dilatatá, subovatá.

Shell suborbicular and depressed, smooth (?), thin, delicate, and fragile, with an involute and very oblique vertex, spirally twisted on the right side, and projecting beyond the base, with an expanded subcircular opening.

Diameter of aperture, $\frac{1}{2}$ an inch.

Locality. Cor. Crag, Sutton.

I have about a dozen specimens of this shell in my cabinet, which, however, are not very perfect, but appear to differ from any of the preceding species sufficiently to be considered distinct ; they are perfectly smooth, without striæ, but they are much injured, and cannot be depended upon in that character. The great obliquity of volution, which much exceeds that of *C. militaris*, is the character which best distinguishes it, and gives it a form approaching to the genus Velutina. Unfortunately, I have no specimen that will show its muscular impression, but the flattened form of the posterior margin, beneath the vertex, shows the capuloid character of the shell ; this obliquity of its volutions gives it a subcarinated form to the left side of the back, and connects this species with the regularly but obliquely spiral univalves.

This genus is largely developed in the Crag formations, but appears to have been subject to great variation. My specimens are here separated and considered as determinable into four species, which are, however, so ill defined, with the line of specific demarcation so obscure, that their correct appropriation is still a matter of doubt.

CALYPTRÆA,* *Lam.* 1801.

PATELLA (spec.) *Linn.*
GALERUS. *Humph.* 1797.
CALYPTRA. Id.
CALYPTRUS. *Montf.* 1810.
INFUNDIBULUM. *J. Sow.* 1815.
TROCHITA. *Schum.* 1817.
MITRULARIA. Id.
MITELLA. *Leach*, 1818.

Gen. Char. Shell conical, clypeiform, depressed, subspiral, with the vertex nearly central; base more or less regularly orbicular, entire, and sharp edged; interior cavity furnished with a lateral appendage or septum, variable in form, externally smooth, rugose, costated, or sometimes covered with imbricated spines.

This genus has been separated by M. Lesson into several subgenera, distinguished by the form of the internal diaphragm or appendage, which is exceedingly variable both in shape and magnitude. Mr. Broderip has figured and described in the Trans. of the Zool. Soc. 1835, a large number of recent species, belonging to the five Sections into which they have been divided by M. Lesson, viz., Calyptræa, Calypeopsis, Syphopatella, Crepipatella, and Crepidula, and says the differences are not such as will entitle them to generic distinction, and quotes Mr. Owen's opinion in regard to the animal, who says: " The soft parts of the animal of Crepidula are the same with those of Calyptræa in all essential points of structure, differing only in the proportionate extent of the anterior part of the foot, and dorsal groove of the mantle." *Dispotea*, Conrad, was intended for those species that have been called cup-and-saucer limpets, in which the internal appendage resembles the subconical form of the external shell.

The known recent species are marine animals, and most of them have been found on rocky coasts in warm climates. M. d'Orbigny describes one species from the Cretaceous formation, which is, I believe, the earliest appearance of this genus.

---

* Etym. Καλύπτρα, a covering.

1. Calyptræa Chinensis. *Linn.* Tab. XVIII, fig. 1, *a—e.*
    Patella Chinensis. *Linn.* Syst. Nat. p. 1257, 1766.
       — Sinensis. *Gmel.* p. 3692.
       — rotunda. *List.* Conch. t. 546, fig. 39.
       — — *Park.* Org. Rem. t. 5, fig. 10, 1811.
       — albida. *Don.* Brit. Shells, pl. 129.
       — Chinensis. *Mont.* Test. Brit. p. 489, t. 13, fig. 4, 1802.
       — muricata. *Brocchi.* Conch. foss. Subapenn. pl. 1, fig. 2, 1814.
    Mitella Sinensis. *Leach,* MS. 1818.
    Calyptrea Chinensis. *Flem.* Brit. An. p. 362, 1828.
    Infundibulum rectum. *J. Sow.* Min. Conch. t. 97, fig. 3, 1825.
       — clypeum. *Woodward.* Geol. of Norf. t. 3, fig. 2, 1833.
    Calyptræa lævigata. *Desh.* 2d edit. Lam. tom. vii, p. 623, 1836.
       — recta. *Nyst.* Coq. foss. de Belg. p. 361, pl. 35, fig. 11, 1844.
       — Sinensis. - - - pl. 35, fig. 14.
       — squamulata. - - - pl. 35, fig. 13.
       — vulgaris. *Phil.* En. Moll. Sic. vol. ii, p. 93, 1844.
       — muricata. *Dujard.* Mém. Soc. Géol. de France, t. 11, pt. 2, p. 274, 1837.
    Infundibulum rotundum. *S. Wood.* Catalogue 1842.
       — subsquamosum. Id.
       — sinense. *Morris.* Cat. of Brit. Foss. p. 148, 1843.
    Calyptrea Chinensis. *Thorpe.* Brit. Mar. Conch. p. 136, 1844.

C. Testá variabili, orbiculari, conoideá, plus minusve depressá, tenui, fragili, læviusculá aut squamulatá; striatá; striis remotiusculis; spiraliterque circinatis; vertice centrali, spirá perspicuá, ad dextram revolutá; lamina interná, simplici incurvá.

Shell variable, with an orbicular base, elevated or depressedly conical, sometimes nearly discoidal, generally smooth in its young state, often with one or more rows of imbricated and elevated spines; vertex central, with a visible spire, and a sharp and arcuated diaphragm, reflected and folded centrally.

Diameter of base, 1½ inch.
Locality. Cor. Crag, Sutton, Gedgrave, Ramsholt.
    Red Crag, *passim.*   Recent, British Seas.

Small specimens of this species are abundant in the Coralline as well as in the Red Crag. In my Catalogue the large squamose and imbricated specimens were supposed to have constituted another species; but on an examination of more specimens, the distinction is lost by the imperceptible gradation and intermingling the one with the other, and the line of demarcation cannot be pointed out. Some specimens are much elevated, like a Chinese Cooly's cap; while others are so depressed as to be almost discoidal. Some are imbricated all over the surface, while others have one, sometimes two, rows of elevated spines.* M. Philippi considers *P. muricata,* Broc., as only a variety, and I am of the same opinion.

---

* Calyptræa Chinensis.—Vestiges of these spines may be detected in the recent British specimens, even in those which do not exceed three eighths of an inch in diameter, although shells of that size are

TECTURA,* *Audouin* and *M. Edwards,* 1830.

PATELLA (spec.) *Linn.*
LOTTIA. *J. E. Gray,* 1833.
ACMÆA. *Esch.* 1833.
PATELLOIDA. *Quoy* and *Gaimard.*
PATELLOIDEA. *Cantraine.*
HELCION. *D'Orb.*
PATELLA (sect.) *Lovén,* 1846.

Gen. Char. Shell symmetrical, conical, patelliform, generally depressed, and rather thin, often subhyaline and smooth, or faintly radiated upon the exterior; vertex eccentric, inclining towards the anterior; base ovate, more or less elongate; muscular impression, of a horse-shoe form, open in front.

This genus has been established entirely upon the animal inhabitant, by which alone it can be distinguished; as in the form of the shell with its radiating marks, a well as in the shape of the muscular impression, it is precisely similar to Patella. The generic position of fossil species can on that account only be determined by their presumed identity with well-known recent shells. The animal, however, is said not to have its branchiæ like those of Patella, and it is now placed in a different ordér. Mr. Gray informed me that his Paper, describing the animal, was read at the Zoological Society in June, 1833, and was published during that year. The name of Acmæa bears also a date of the same year, but Tectura, on the authority of that gentleman, was published in 1830, and of course has the precedence.

The shells of this genus are generally thinner, more fragile, and more finely striated than those of Patella; but these characters are variable and insufficient for generic determination. M. d'Orbigny is of opinion that the Patelliform shells of the older rocks were animals belonging to this genus, but the reason for this opinion is not given.

considered as smooth, with only the visible lines of growth. The diaphragm is largely reflected over the columella, and the outer edge sharp and arched, extending within a short distance of the margin of the shell. This is a dextral shell, and regularly spiral. Specimens of this species, from the Red Crag, are completely enveloped externally, with an agglomerated mass of cellepora; a probable incrustation when the animal was living, as in no instance has it extended beyond the margin into the interior, which is perfectly clear.

* Etym. *Tectura,* a covering.

1. TECTURA VIRGINEA. *Müll.* Tab. XVIII, fig. 6, *a—c.*

PATELLA VIRGINEA. *Müll.* Zool. Dan. p. 13, pl. 12, fig. 4-5, 1773.
— PARVA. *Da Costa.* Brit. Conch. t. 8, fig. 11, 1778.
— — *Mont.* Test. Brit. p. 480, 1803.
— EQUALIS. *Sow.* Min. Conch. t. 139, var. β, 1816.
— — *Nyst.* Coq. foss. de Belg. p. 349. pl. 35, fig. 5, 1844.
var. β, CONICA. *Jeffreys.*
PATELLA PULCHELLA. *Forbes* ex *Lovén.* 1846.
LOTTIA VIRGINEA. *S. Wood.* Catalogue 1842.
ACMÆA VIRGINEA. *Thorpe.* Brit. Mar. Conch. p. xxxi, 1844.

T. Testá tenui, clypeiformi, conicá, depressá vel elevatá; suborbiculari vel ovatá; radiis obsoletis; vertice obtuso; margine integro.

Shell shield-shaped, with an ovate base, elevated or depressed, externally radiated; radii nearly obsolete; vertex obtuse, eccentric; margin smooth.

Longest diameter of base, $\frac{5}{8}$ of an inch.

Locality. Red Crag, Sutton, Bawdsey, and Brightwell. Recent, Britain.

There are two varieties of this species, neither of which is very rare: var. *a* (fig. *a*) is ovate, longer than broad, and depressed, with a very eccentric apex; var. β (fig. *c*), *P. conica,* Jeff., has a subcircular base, and a more elevated vertex.

The muscular impression in this shell is precisely similar to that in Patella, and, from the shell alone, could not be distinguished. The apex in all my specimens is more or less eroded. In var. *a,* the vertex is not more than a quarter of the longest diameter from the anterior margin, with an elevation of scarcely one third; in var. β, the vertex is equally eccentric, but the base is less ovate, and its height is equal to two thirds of the longest diameter.

2. TECTURA FULVA. *Müller.* Tab. XVIII, fig. 7, *a—b.*

PATELLA FULVA. *Müller.* Zool. Dan. t. 24, fig. 1-3, 1773.
— FORBESII. *Smith.* Wern. Mem. vol. viii, p. 59, pl. 2, fig. 3, 1838.
— ACUMINATA (?). *Grateloup.* Bord. Foss. pl. 1, fig. 8-10, 1837.
CALYPTRÆA (?) DUBOIS. *De Montp.* pl. 4, fig. 10-11, 1831.
PATELLA FULVA. *Lovén.* Ind. Moll. Scand. p. 26, 1846.

T. Testá minutá, tenui, ovatá, depressá, striatá; striis longitudinalibus regularibus, tuberculatis, vel granulatis: vertice eccentrico, acutiusculo.

Shell small, and thin, with an ovate base, depressedly conical, covered with regular tuberculated or granulated rays; vertex eccentric, and rather sharp.

Longitudinal diameter, $\frac{5}{16}$; *transverse diameter,* $\frac{4}{16}$; *altitude,* $\frac{3}{16}$ of an inch.

Locality. Cor. Crag, Sutton. Recent, British and North Seas.

This is by no means a rare shell in the Coralline Crag. The form of the base is ovate, rather narrower on the anterior portion, with a very eccentric vertex; the shell is generally depressed, although some specimens are somewhat elevated; and the vertex is generally eroded; when perfect, it is rather acute.　The rays are regular in distance upon the same specimen, though very variable in number.　A recent specimen, obligingly sent to me by William Thompson, Esq., for comparison, is perfectly ovate in the base, with more numerous rays than in the generality of my specimens, which are, however, rather variable in that character; when perfect, the tuberculated rays will distinguish it, and may be considered a fair identification.

Some small specimens from the Red Crag, which had been considered as the young of *T. virginea*, may possibly belong to this species, as they are the same in form, but, having lost the tubercles or granules from the rays, cannot be distinguished.　The specimen sent by Mr. Thompson was dredged in deep water off Cape Clear by Mr. M'Andrew.

3. TECTURA (?) PARVULA.　*Woodward.* Tab. XVIII, fig. 8, *a—b.*

LOTTIA PARVULA.　*S. Wood.* Catalogue 1842.

PATELLA PARVULA.　*Woodward.* Geol. of Norf. t. 3, fig. 1, 1833.

T. Testá oblongo-ovatá, elongatá, costatá vel radiatá; interstitiis, longitudinaliter tenuissimè striatis; vertice obtuso, eccentrico.

Shell oblong-ovate, elongate, costated, or rather radiated, very finely striated between the rays, with an obtuse vertex before the centre of the shell.

Length, $\frac{1}{8}$; *breadth,* $\frac{1}{10}$ of an inch.

Locality.　Mam. Crag, Bramerton.

This shell appears to be rare.　My cabinet contains but one specimen, which, with two others belonging to Mr. Wigham, are all I have seen; it agrees with Woodward's figure, but I am not acquainted with any recent form that it precisely resembles; it differs from either of the other species in the fewer number and subangular form of its rays as well as in the more oblong form of its base.　Between the rays the shell is covered with very fine longitudinal striæ, and the margin is crenulated.　This may possibly be the young of an elongated variety of *Patella vulgata*.

SCISSURELLA,* *D'Orbigny,* 1823.

Gen. Char. Shell small, thin, and fragile, subglobose, with, generally, a depressed spire; aperture subcircular or ovate, modified by the elevation or depression of the spire; peritreme sharp, with a deep sinus or slit, sometimes an oblong foramen in the outer lip, with a sort of double keel on the outside formed by the reflected edge of the sinus; base of the shell with a generally large, open, and deep umbilicus.

The animal of this genus is at present unknown, and its position is, consequently,

* Etym. *Scissus,* cut, divided, *a scindo.*

uncertain; the sinus in the outer lip resembles that of Pleurotomaria. Mr. Gray has placed it among the Haliotidæ.

1. SCISSURELLA CRISPATA. *Flem.* Tab. XV, fig. 13.

      SCISSURELLA CRISPATA. *Flem.* Hist. Brit. Animals, p. 366, 1828.
         —   —   *S. Wood.* Catalogue 1842.
         —   ASPERA (?). *Phil.* En. Moll. Sic. p. 160, t. 25, fig. 17, 1844.

Sc. Testá orbiculato-ovatá, heliciformi, vertice depresso; anfractibus tribus, convexis; in medio cingulis duobus transversis instructis; longitudinaliter plicatis, plicis elegantissimè curvatis, interstitiis transversim striatis.

Shell orbiculato-ovate, with a depressed apex; volutions three, convex, and rather tumid, with a double keel in the middle or outermost portion of the volution; longitudinally plicated, with elegantly curved and crispated ridges, between which it is spirally striated.

Diameter, $\frac{1}{20}$ of an inch.

Locality. Cor. Crag, Sutton.   Recent, Scottish Coast.

I have unfortunately but one slightly injured specimen, having a part of the outer lip destroyed; however, what remains corresponds so well with Dr. Fleming's clear and copious description, that I have no hesitation in assigning it to his species. It appears to be rather less elevated in the spire than *Sc. aspira,* Phil. (En. Moll. Sic. 1844, vol. ii, p. 160, t. 25, f. 17), which it otherwise much resembles.*

EMARGINULA, *Lam.* 1801.

    EMARGINULUS. *Montf.* 1810.
    SUBEMARGINULA. *Blainv.* 1825.
    HEMITOMA. *Swains.* 1840.
    CLYPIDINA. *Gray,* 1847.

Gen. Char. Patelliform, conical, or clypeiform; vertex more or less elevated, and curving slightly backwards; aperture elongato-oval; anterior margin with a fissure, thickened within; exterior generally raised and decussated; muscular impression open in front.

Several shells, possessing this patelliform character, with a fissure in front for the passage of the water to the branchiæ, but differently situated, were placed by Lamarck

* Since the above was written I regret to say that the specimen, while in the hands of the engraver, was lost. The figure above referred to is, I believe, a faithful representation of what it was.

in this genus; they have recently been separated into distinct genera, according to the position of this aperture. In this genus the margin is cut by the fissure; in Rimula the slit is removed into the middle of the anterior portion, leaving the margin entire; in Cemoria the aperture is in front of the vertex on the top, thereby connecting it with Fissurella, from the young of which it is scarcely to be distinguished. The muscular impression nearly surrounds the shell a little above the margin, leaving an opening in front for the head of the animal. As the shell is enlarged, this muscular mark is removed and probably covered with fresh calcareous matter, as this necessarily varying impression is nowhere to be seen in the upper or younger portion of the shell. The genus made its first appearance, as far as is yet known, in the Oolitic period.

1. EMARGINULA FISSURA. *Linn*. Tab. XVIII, fig. 3, *a—b*.

PATELLA FISSURA. *Linn*. Syst. Nat. p. 1261.
— *Müller*. Zool. Dan. t. 24, fig. 1-3, 1773.
EMARGINULA RETICULATA. *J. Sow*. Min. Conch. t. 74, 1815.
— — *Nyst*. Nouv. rech. Coq. foss. Prov. d'Anv. p. 16.
— FISSURA. *Dujard*. Mém. Soc. Géol. de France, tom. ii, p. 273, 1837.
— — *Nyst*. Coq. foss. de Belg. p. 350, pl. 35, fig. 6, 1844.
— — *S. Wood*. Catalogue 1842, var. *a*, VULGARIS, fig. 3, *a*.
— — - - - var. *β*, PUNCTURA, fig. 3, *b*.
— ROSEA. *Thorpe*. Brit. Mar. Conch. p. 133, fig. 79, 1844.

E. Testá elevatá, conicá, costellatá; costellis longitudinalibus, striisque transversis cancellatá; vertice recurvo; aperturá oblongo-ovatá; margine crenulatá.

Shell elevated and conical; with the vertex slightly recurved; longitudinally costated, and transversely decussated; costæ generally alternating, one large and one small; base of shell oblongo-ovate, with a crenulated margin.

Longitudinal diameter of base $\frac{1}{2}$; *Altitude and breadth*, $\frac{3}{8}$ of an inch.

Locality. Cor. Crag, Sutton.

Red Crag, *passim*. Recent, Britain.

An abundant shell in both formations. In the young state the recurved vertex is distinctly visible; in large specimens it is not seen. The shell is very variable in its proportions, some specimens not having a height more than half the longest diameter, while, in others, it is even higher than long: the vertex, in some is nearly central, and in others it hangs as far back as the basal margin, and even beyond it. The exterior is regularly decussated, leaving large and deep alveoli upon the surface; the fissure is narrow, thickened internally, especially round the upper part, and in length about one third of the height. What I had considered as a distinct species, and published in my Catalogue under the name of *punctura*, is probably only an extreme variety, with a worn and altered surface. In this shell (fig. 3, *b*,) the margin does not appear to have been periodically thickened, so that the longitudinal costæ, which are 20—22, stand prominently out, and the shell appears striated with small punctures between them; these small holes are probably the alveoli left by the decussating margin. In most of the specimens of the more common variety the costæ are all equally prominent, and decussated by the thickened margin; in others, these costæ appear to alternate, one large and one small. In this variety there are from three to five striæ or costulæ between the more elevated ribs, which are also rounded and smooth. In the common variety the costæ are made rough and nodulous by the thickened margin of the shell.

Specimens of this species are occasionally loaded with a mass of cellepora.

In some specimens the rays are all equally prominent, amounting to upwards of sixty, but in the greater number there is a smaller intermediate one.

2. EMARGINULA CRASSA. *J. Sow.* Tab. XVIII, fig. 2, *a—e.*

EMARGINULA CRASSA. *J. Sow.* Min. Conch. t. 33, 1813.
— *Nyst.* Coq. foss. de Belg. p. 352, pl. 36, fig. 3, 1844.
— *Forbes.* An. and Mag. Nat. Hist. vol. xiv, p. 410, pl. 10, fig. 1, 1844.

E. Testá conicá, depressá, vel convexá, costatá; costis tenuibus, radiantibus, crebris; striis transversis granulosis decussatis; aperturá ovatá; margine crenulatá; rimá valdè excavatá.

Shell depressedly conical and striated, with the anterior dorsal portion convex; longitudinally striated and transversely decussated; base ovate; margin crenulated; fissure large and deep.

Longitudinal diameter, 2 inches.

Locality. Cor. Crag, Sutton and Ramsholt.

Red Crag, Sutton and Newbourn. Recent, British Seas.

I have only a few specimens of this shell from either formation. According to Professor E. Forbes, the species is found living in the Scandinavian seas, and has recently been dredged up alive by Mr. M'Andrew off the Scottish coast. The fissure is rather wide, and in depth about one fifth of the entire length of the anterior portion, and the breadth of the base is less before than it is behind the vertex, which is itself variable in position, as may be seen in the figures above referred to. The longest diameter of my largest specimen measures two inches, and is from the Red Crag, at Newbourn; but a larger individual has, I am informed, been obtained by W. H. Alexander, Esq. of Ipswich.

Figs. *c* and *e* are from the Coralline Crag, Ramsholt.

CEMORIA,* *Leach*, 1819.
DIADORA. *Gray*, 1821.
SYPHO. *Brown*, 1827.
PUNCTURELLA. *Lowe*, 1828.
RIMULA. *Lovén*, 1846.
FISSURELLA (spec.) *Flem.*
PATELLA (spec.) *Müll.*

Gen. Char. Shell small, patelliform, like an Emarginula, with an ovate base, and entire margin; apex elevated, slightly recurved; a perforation in the upper part of the shell before the vertex, and a thickened plate within, above the fissure.

But one species of this genus is as yet known, and this has been placed with Fissurella, from which, however, it is now determined to be perfectly distinct. It much resembles the genus Rimula, in having its opening between the margin and the vertex, and rather nearer to the latter; but a recent species of the genus Rimula, in the British Museum, shown me by J. E. Gray, Esq., has a fissure in the centre of the shell, without the internal thickened plate of the present genus; but until more information is obtained respecting the animal inhabitant, this may be considered as a distinct genus.

1. CEMORIA NOACHINA. *Linn.* Tab. XVIII, fig. 5, *a—c.*
PATELLA NOACHINA. *Linn.* Mantissa, p. 551.
—— FISSURELLA. *Müll.* Zool. Dan. i, t. 24, fig. 4-6.
FISSURELLA NOACHINA. *Lyell.* Obs. upon the Elev. of Sweden.
PUNCTURELLA NOACHINA. *Lowe.* Zool. Journ. vol. iii, p. 77, 1828.
SYPHO STRIATA. *Brown.* Illust. Brit. Conch. pl. 36, fig. 14-16, 1827.
CEMORIA FLEMINGII. *Leach.* MS. 1819.
—— NOACHINA. *Gould.* Inv. of Massachus. p. 156, fig. 18, 1841.

C. Testá conicá, crassá, costatá, costis rugosis, confertis, inæqualibus, radiantibus; vertice vix intorto; basi ovatá; margine crenulatá.

Shell strong, conical, and costated, with numerous rugose rays or costæ, generally alternating large and small, with the fissure of a triangular shape, and a deep muscular impression.

Longest diameter of base, ⅝; *altitude,* ⅜ of an inch.

Locality. Mam. Crag, Bridlington.

Recent, Arctic Seas, and Seas of Boreal America.

This is at present known as a Crag fossil only from the above locality. The specimen figured is from the cabinet of Mr. Leckenby. One specimen, also, is among the shells sent by Mr. Bean. There is, I conceive, but little doubt of its identity with

* Etym. (?)

the recent species. The form of the branchial aperture is nearly triangular, with the base upwards, the point extending half way to the margin, with a thickened ridge or plate within, above the fissure, and a muscular band extends round the posterior half of the shell, about one third of its height from the margin; this muscular band terminates in an extended angular-shaped impression on each side. The form of the base is somewhat of an oblongo-ovate form, the posterior portion being rather the broader.

Dr. Gould mentions this species as frequently taken in the stomachs of fishes upon the coast of Massachusetts; and has given, as a synonyme, the *Patella apertura*, Mont.; there is, I believe, no doubt of that shell being the young of *Fissurella cancellata*.

FISSURELLA,* *Bruguière*, 1789.

LARVA. *Humph.* 1797.

FISSURELLUS. *Montf.* 1810.

LUCAPINA. *Gray.*

PUPILLÆA. Id.

MACROCHISMA. *Swains.* 1840.

FISSURIDEA. Id.

CLYPIDELLA. Id.

FISSURELLIDIA. *D'Orb.*

Gen. Char. Shell depressedly conical, with an ovate base; an aperture in the vertex, on the shorter side, towards the head of the animal; generally ovate, or of a keyhole shape, sometimes contracted in the centre, sometimes tripartite; rayed or costated upon the exterior, and often cancellated by the thickened edges of the different periods of growth; margin generally thickened; muscular impression large, nearly surrounding the interior.

This genus differs from Emarginula in having the perforation at the summit, through which the water is conveyed to the branchiæ; the opening is on the shorter or anterior part of the shell; but in the very young state this aperture is nearly central, with a small involute vertex, curving towards the posterior; and as the animal

increases in size, it has the power of absorbing or dissolving a portion of the shell, and the aperture is thereby made larger, and thickened round the edge; the recurved vertex is then lost, and in the full-grown shell no trace of it remains. From this great difference in appearance between the young and old shells, Montague has made a distinct species of the young state of our Crag shell, which he has named and described as *Patella apertura*.

---

* Etym. The diminutive of *Fissura*, a fissure or cleft.

1. FISSURELLA GRÆCA. *Linn*. Tab. XVIII, fig. 4, *a—c*.

    PATELLA GRÆCA.   *Linn*. Syst. Nat. p. 1262, 1766.

      —   CANCELLATA.   *Lister*. t. 527, fig. 2, 1685.

      —   GRÆCA.   *Mont*. Test. Brit. p. 492, 1803.

      —   APERTURA.   -   -   p. 491, t. 13, fig. 10.

      —   RETICULATA.   *Don*. Brit. Shells, t. 21, fig. 3.

    FISSURELLA GRÆCA.   *J. Sow*. Min. Conch. t. 483, 1826.

      —   *Brown*. Illust. Brit. Conch. pl. 36, fig. 10-11, 1827.

    SYPHO RADIATA.   -   -   -   pl. 36, fig. 20.

    FISSURELLA GRÆCA.   *Flem*. Brit. An. p. 364, No. 362, 1828.

      —   APERTURA.   -   -   No. 363.

      —   GRÆCA.   *Nyst*. Coq. foss. de Belg. p. 353, pl. 35, fig. 7, 1844.

      —   CANCELLATA.   *S. Wood*. Catalogue 1842.

      —   —   *Morris*. Catalogue of Brit. Fossils, 1843.

F. Testá clypeiformi, oblongo-ovatá, depresso-conicá; costis valdè elevatis, lineisque transversis cancellatis, ad intersectiones tuberculatis; foramine ovato; margine crenulatá.

Shell shield-shaped, or oblongo-ovate, covered with elevated rounded rays, decussated by the thickened or slightly reflected lines of growth; aperture oval, thickened within, and rather contracted in the middle; vertex truncated, or sloping towards the anterior margin; margin slightly thickened and crenulated.

Longest diameter, 1½ inch.

Locality.   Cor. Crag, Sutton and Ramsholt.

       Red Crag, Sutton, Bawdsey, Newbourn, and Walton.

Recent, British Seas.

Young specimens of this shell are abundant in the Coralline Crag, in which the involute vertex is remaining, but I have not seen this inflected portion on any shell which has attained the length of one quarter of an inch. This species is subject to much variation in form, but the posterior portion is always broader, as well as longer, than the anterior; sometimes the shell is a good deal elevated, and the sides of the cone convex; in others it is depressed, with the sides flat, inclining to concavity, from the vertex to the margin, which is a little thickened and crenulated. The rays generally alternate, one large and one small, though sometimes they are placed in sets, each fifth one being more elevated than the others. Fig. 4 *c* is the young state of this shell. This is given as an Eocene fossil, by M. Deshayes, from Grignon.

Ord. *TECTIBRANCHIATA.*

ACTÆON,* *Montf.* 1810.
VOLUTA (spec.) *Linn.*
SOLIDULA. *Fischer.*
TORNATELLA. *Lam.* 1816.

Gen. Char. Shell ovate, convolute, subcylindrical, with a more or less elevated spire; striated or sulcated in a transverse or spiral direction; aperture linear, narrow above, rounded below; outer lip plain, generally sharp; columella spiral, with tooth-like projection; operculum corneous.

1. ACTÆON NOÆ. *J. Sow.* Tab. XIX, fig. 6, *a—b.*
ACTÆON NOÆ. *J. Sow.* Min. Conch. t. 374, 1822.
— *S. Wood.* Catalogue 1842.
TORNATELLA NOÆ. *Nyst.* Coq. foss. de Belg. p. 424, pl. 37, fig. 22, 1844.

A. Testá ovato-elongatá vel subcylindraceá; spirá conoideá; anfractibus septem vel octo convexiusculis, transversim striatis; suturis depressis, subcanaliculatis; columellá uniplicatá, labro acuto, intus sulcato.

Shell elongato-ovate, or subcylindrical, with a conical spire; whorls slightly convex, regularly striated, or rather sulcated, transversely; suture deep, subcanaliculated; aperture linear; outer lip sharp, with the striæ visible within.

Axis, 1 inch; *diameter*, $\frac{9}{16}$.

Locality. Red Crag, Walton Naze and Brightwell.

This species is not very rare at the above locality, though, from its extreme fragility, it is difficult to obtain in perfect condition. The shell appears to have undergone a slight diminution in magnitude, and to have shrunk from the inner lip, perhaps from the loss of a more than ordinary quantity of animal matter. It is considered here as differing from *A. tornatilis* in having the spire rather shorter, and a deeper and more distinct canal at the suture, which, however, might be produced by the shrinking of the shell; but the upper part of the volution is more rounded, and the striæ are regularly distributed over the surface, sometimes single, sometimes in pairs. In the recent British shell they are larger and coarser at the base, and the fold upon the columella is smaller. The deep sulci of the exterior are visible within at all ages from the thinness of the shell. Aperture three fourths of the length of axis.

* Etym. *Actæon*, a proper name.

2. ACTÆON TORNATILIS. *Linn.* Tab. XIX, fig. 5, *a—b.*

VOLUTA TORNATILIS. *Linn.* Syst. Nat. p. 1187, 1758.

— *Mont.* Test. Brit. p. 231, 1803.

TORNATELLA FASCIATA. Ency. Méth. pl. 452, fig. 3, 1816.

ACTÆON STRIATUS. *J. Sow.* Min. Conch. t. 460, fig. 2, 1824.

TORNATELLA TORNATILIS. *Flem.* Brit. An. p. 336, 1828.

— FASCIATA. *Phil.* En. Moll. Sic. vol. i, p. 166, 1836.

ACTÆON TORNATILIS. *S. Wood.* Catalogue 1842.

— *Morris.* Cat. Brit. Foss. p. 138, 1843.

TORNATELLA FASCIATA. *Desh.* 2d edit. *Lam.* ix, p. 41, 1843.

— STRIATA. *Nyst.* Coq. foss. de Belg. p. 426, pl. 37, fig. 24, 1844.

— TORNATILIS. *Phil.* En. Moll. Sic. vol. ii, p. 143, 1844.

A. Testá ovato-cylindraceá; spirá conoideá; apice acuto; anfractibus sex convexiusculis; transversim striatis; columellá uniplicatá; labro acuto, intus non sulcato.

Shell ovato-cylindrical, with a slightly elevated spire, and acute apex; volutions six, slightly convex; finely striated on the body of the shell, more coarsely so at the base; columella with one fold; outer lip sharp, ridges not visible on the inside.

Axis, $\frac{5}{8}$; *diameter,* $\frac{5}{16}$ of an inch.

Locality. Cor. Crag, Sutton.

Red Crag, Sutton. Recent, British Seas.

Of this species I have found but very few specimens, from either the Red or Coralline Crag. It appears to be distinct from the preceding, and I do not perceive any sufficient character by which it can be separated from the recent British species. The upper part of the volution is flatter than in *A. noæ*, the aperture being sharply acuminated at the junction of the outer lip with the body whorl, and consequently it has not so high a shoulder. Mr. Sowerby must have had an imperfect specimen for his figure, as, in his description, he says "columella without a plait;" in my specimens the fold is very distinct, though less prominent than in *A. noæ*. Aperture three fifths of the length of the axis.

3. ACTÆON SUBULATUS. *S. Wood.* Tab. XIX, fig. 7, *a—b.*

ACTÆON SUBULATUS. *S. Wood.* Catalogue 1842.

A. Testá elongato-ovatá, subulatá; spirá elevatá, apice acuto; anfractibus septem convexiusculis, subdepressis, transversim tenuissimè striatis; aperturá ovato-angustá; columellá uniplicatá.

Shell elongato-ovate, tapering, with an elevated spire, and acute apex; whorls seven, slightly convex, somewhat depressed, transversely striated; striæ broader and coarser at the base; outer lip thin, with an obtuse fold upon the columella.

Axis, $\frac{5}{8}$ of an inch.

Locality. Red Crag, Sutton.

This is not so rare as the preceding species, from which it appears to differ, in being more elongated, with a more acuminated apex, and with the volutions less convex. The aperture in this measures about half the entire length of the shell, but that is a character liable to vary, and not alone to be depended upon, as those proportions would alter by the simple elevation or depression of the spire. This species has a more attenuated form, and the plication or fold upon the columella is less prominent than that of the preceding.

4. ACTÆON LEVIDENSIS. *S. Wood.* Tab. XIX, fig. 4, *a—b.*
        MELANIA AURICULA (?). *Grat.* Bord. foss. p. 7, pl. 5, fig. 4, 1838.
        ACTÆON LEVIDENSIS. *S. Wood.* Catalogue 1842.
        TORNATELLA ELONGATA (?). *Nyst.* Coq. foss. de Belg. p. 426, pl. 37, fig. 23, 1844.

A. Testá minutá, turrito-cylindricá, spirá elevatá; apice obtuso; anfractibus quinque convexiusculis, transversim regulariter sulcatis; sulcis imbricatis; aperturá angustá; columellá uniplicatá.

Shell small, tapering, and elongate, subcylindrical, thick and strong, with an elevated spire, and obtuse apex; whorls five, slightly convex; striated or sulcated transversely; suture distinct and deep, with an obtuse fold upon the columella; outer lip sharp, slightly curved.

Axis, $\frac{1}{7}$ of an inch.

Locality. Cor. Crag, Sutton.

I have found fifty specimens of this species, nearly all of the same size, and presume it to be a full-grown shell. It is covered with elevated transverse striæ, or rather ridges, amounting to about twelve or fourteen, in the last volution, the furrows between which are decussated by elevated lines of growth, and it has an obtuse fold upon the columella. *Melania auricula,* Grateloup, may possibly be the same as our shell, but the figure by that author represents his species as much larger, and without a fold upon the columella; and his description, " transversim subtillissimè striata," does not correspond with our shell, in which the striæ or ridges are very broad, flat on the top, and smooth. M. Nyst, at p. 426, speaks of *Tornatella elongata* as an English Crag shell; I have carefully compared my Barton specimens of what I consider to be Sowerby's *Actæon elongatus,* and they are, I think, quite distinct from my Crag specimens, and Mr. Edwards could not identify any of his Eocene shells with my Crag species.

Both the Crag and Barton species were sent by myself to the Count du Chastel some years since, and a mistake respecting the localities may have occurred.

BULLA,* *Klein*, 1753.

ATYS.   *Montf.* 1810.
BULLUS.   Id.
RHIZORUS.   Id.
SCAPHANDER.   Id.
APLUSTRUM.   *Schum.* 1817.
ASSULA.   Id.
HYDATINA.   Id.
NAUCUM.   Id.
DIAPHANA.   *Brown,* 1827.
RETUSA.   Id.
VOLVARIA.   Id.
CYLINDRELLA.   *Swains.* 1840.
BULLINA.   *Ferrusac.*
CYLICHNA.   *Lovén.* 1846.
AMPHYSPHYRA.   Id.
ALICULA.   *Ehrenb.* (ex *Lovén.*)

Gen. Char. Shell ovate or cylindrical, the last whorl sometimes entirely enveloping the preceding ones; convulute upon a nearly horizontal axis, with a spire generally visible, but always short; aperture the length of the shell, effuse at the lower part, and narrow above; often striated or ornamented on the exterior, sometimes smooth and glossy; texture thick, opaque, corneous or subhyaline; outer lip sharp and smooth.

Shells belonging to this genus, but possessing a visible and slightly elevated spire, have been separated from it and united into a genus, under the name Bullina; such characters, depending merely upon the deviation from the horizontal form of volution, or rather upon the elevation or depression of the outer lip, are not satisfactory, and the gradation from the visible and elevated spire, to the truly spindle-shaped form produced by the elevation and acumination of the upper part of the outer lip is imperceptible.

The animal, it is said, is remarkable for not having antennæ, and for the foot being laterally expanded, so as to be used in the way of fins, to swim with, and the mantle does not include the shell.

Many of the species are found in shallow water, while others have a considerable range in depth, and some are frequenters of estuaries.

Mr. Bean appears to have found this genus in the cornbrash, and one is also figured in the 'Transactions of the Geological Society' as from the Wealden formation. Numerous species are known in the Eocene period, and it is widely extended as a recent genus, being found in all parts of the world, and not restricted to any climate. Those found in the Crag, identical with existing species, are either British or Northern forms.

* Etym. A bubble.

1. BULLA LIGNARIA. *Linn.* Tab. XXI, fig. 8, *a—b.*

    BULLA LIGNARIA. *Linn.* Syst. Nat. p. 1184, No. 379, 1766.
        —   *Donovan.* Brit. Shell, t. 27, 1807.
        —   *Mont.* Test. Brit. p. 205, 1803.
    SCAPHANDER LIGNARIUS. *Montf.* Conch. Syst. p. 334, 1810.
        —   LIGNARIA. *Leach.* Moll. 1820.
        —   LIGNARIUS. *Lovén.* Ind. Moll. Scan. p. 10, 1846.
    BULLA LIGNARIA. *Brown.* Illust. Brit. Conch. pl. 38, fig. 23-24, 1827.
        —   *Broc.* Conch. foss. Subapenn. t. 2, p. 274.
        —   *Dujardin.* Mém. Soc. Géol. de France, tom. 11, pl. 2, p. 275, 1837.
        —   *Phil.* En Moll. Sic. vol. ii, p. 95, 1844.
        —   *S. Wood.* Illust. Mag. Nat. Hist. pl. 7, fig. 14, 1839.
        —   *Nyst.* Coq. foss. de Belg. p. 452, pl. 39, fig. 5, 1844.
        —   Enc. Method. pl. 359, fig. 3, A.B.

B. Testá ovatá, tenui, laxè convolutá, versus spiram attenuatá, transversim striatá, spirá truncatá, depressá; labro arcuato; sine umbilico.

Shell ovate and thin, slightly convolute, rather contracted towards the spire, with a large expanded aperture, ovate, rounded at the lower part; transversely or spirally striated upon the exterior; striæ distant, impressed; vertex truncate, with a slightly depressed spire.

Axis, 1 inch.

Locality.   Cor. Crag, Sutton.

        Red Crag, Sutton.   Recent, British Seas and Mediterranean.

I have been able to obtain only a few specimens of this shell, which appears to be identical with the well-known recent British species. M. Nyst, in the work above referred to, has given as a synonyme, *B. lignaria,* Deshayes (Coq. foss. des Env. de Paris, pl. 5, f. 4-6), thereby considering this as an Eocene species. I am not acquainted with the Paris basin-shell, but a species somewhat resembling this, from Barton, in the cabinet of Mr. Edwards, and which I imagine to be the same, differs in several particulars; it is more elongate, less expanded, more closely and regularly striated, with the striæ beautifully punctured, and is, I think, a distinct species, probably *Bulla Fortisii,* Brongniart.

2. BULLA CONULUS. *Desh.* Tab. XXI, fig. 2, *a—c.*

    BULLA CONULUS. *Desh.* Coq. foss. des Env. de Par. p. 41, pl. 5, fig. 34-36, 1837.
      —   —   *Grat.* Foss. de Bord. p. 51, pl. 3, fig. 4-5, 1838.
      — ANGISTOMA (?). *Grat.*   -   p. 52, fig. 6-7.
      — OVULATA (?). *Dubois.* Foss. de Volhyn. p. 49, pl. 1, fig. 13-14, 1831.
      — CONULUS. *S. Wood.* Illust. in Mag. Nat. Hist. pl. 7, fig. 6, 1839.

B. Testá minutá, ovato-conicá, politá; transversim tenuissimè striatá; spirá occultá; vertice infundibuliformi; aperturá supernè angustissimá; basi dilatatá; columellá infernè obsoletè uniplicatá.

Shell small, ovato-conical, covered externally with fine and distant transverse striæ impressed upon a glossy surface; spire hidden; vertex umbilicated or funnel-shaped; aperture narrow above, and expanded below; left lip slightly replicate, with an obscure fold upon the columella.

Axis, $\frac{1}{4}$ of an inch; *diameter*, $\frac{1}{8}$.

Locality. Cor. Crag, Sutton.

This is by no means a rare species at the above locality. It differs a little, as might be expected, from the Paris basin-shell but not sufficiently to entitle it to specific distinction.

In perfect specimens the striæ may be observed over the whole surface, though most distinct upon the base of the volution; when seen, they appear fewer and more distant than upon the French Eocene specimens, which are rather more elongate or less conical, with a rounded vertex. In the Crag shell the upper part of the lip is rather sharp, producing a funnel-shaped depression at the spire.

3. BULLA ACUMINATA.   *Brug.* Tab. XXI, fig. 7, *a—c.*

    BULLA ACUMINATA.   *Brug.* Enc. Méth. Vers. tom. 1, p. 376, 1789.

    —   *Phil.* En. Moll. Sic. vol. i, p. 122, t. 7, fig. 18, 1836.

    —   *Phil.*   -   -   vol. ii, p. 96, 1844.

    —   *Grat.* Not. sur les Fam. des Bull. p. 63, p^l. 3, fig. 43-44, 1837.

    —   *S. Wood.* Catalogue 1842.

    —   *Nyst.* Coq. foss. de Belg. p. 457, t. 39, fig. 11, 1844.

  CYLICHNA ACUMINATA.   *Lovén.* Ind Moll. Scand. p. 10, 1846.

    (Not BULLA ACUMINATA of Min. Conch.)

B. Testá minutá, subfusiformi, supernè parum attenuatá, sed acuminatá, subtus rotundatá; in medio lævissimá ad basim striatá; aperturá angustè lineari, infernè perparum latiore; columellá basi obsoletè uniplicatá.

Shell small, subfusiform; outer lip contracted, and acuminated at the upper extremity, rounded at the lower part, smooth and glossy in the middle, finely striated at the extremities; striæ thicker, and more distinct around the base; aperture linear, expanded slightly below, with an obsolete fold upon the columella.

Axis, $\frac{1}{8}$ of an inch; *diameter*, $\frac{1}{2}$ its length.

Locality. Cor. Crag, Sutton.   Recent, Mediterranean and British Seas.

An abundant species in the Coralline Crag, but I have found it only in one locality. In my Catalogue this name was given with doubt, as in the descriptions at the above references, it is spoken of as " lævissima," and the figures do not well represent our shell. Mr. Alder, however, informs me that he has lately obtained a specimen in Lamlash Bay, and that Mr. Barlee got one also from Loch Fyne, both of which are striated; the striæ upon the Mediterranean shell have, in all probability, been overlooked, as they are present in the British specimens. This shell being convolute upon a nearly horizontal axis, the spire is of course hidden, the outer lip being elevated, and so contracted at the upper part as to give it the appearance of having an acuminated apex, and instead of approaching Bullina it might be said to form a passage into Ovula, through

Ov. patula, which has a non-reflected outer lip. The shell differs from the general character of this genus only in the acuminated form of the upper part of the outer lip, with the fold at the base, as in Bulla, and not above, as in Ovula, and it is not regularly bicanaliculated. The striæ are most distinct upon the base of the volution, but there are also a few visible at the upper part, whilst, upon the body of the shell, at least in my specimens, it is smooth and glossy.*

4. BULLA CYLINDRACEA. *Penn.* Tab. XXI, fig. 1, *a—c.*

BULLA CYLINDRACEA.   *Pennant.* Brit. Zool. vol. iv, t. 70, fig. 85, 1776.

—   *Mont.* Test. Brit. p. 221, t. 7, fig. 2, 1802.

—   *Don.* Brit. Shells, vol. iv, t. 120, fig. 2, 1805.

BULLA CONVOLUTA.   *Broc.* Coq. foss. Subapenn. t. i, fig. 7, 1814.

—   *J. Sow.* Min. Conch. t. 464, fig. 1, 1824.

VOLVARIA CYLINDRICA.   *Brown.* Illust. Brit. Conch. pl. 38, fig. 36-37, 1827.

BULLA CONVOLUTA.   *Grat.* Not. sur la Fam. des Bull. pl. 3, fig. 37, 1837.

—   *Phil.* En. Moll. Sic. vol. ii, p. 97, 1844.

—   *Nyst.* Coq. foss. de Belg. p. 454, pl. 39, fig. 6, 1844.

BULLA CYLINDRACEA.   *S. Wood.* Illust. in Mag. Nat. Hist. pl. 7, fig. 8, 1839.

CYLINDRELLA ALBA.   *Swains.* Treat. on Malac. p. 326, fig. 94 *b*, 1840.

CYLICHNA CYLINDRACEA.   *Lovén.* Ind. Moll. Scand. p. 10, 1846.

B. *Testá cylindricá crassá; tenuissimè striatá; aperturá lineari, versus basim parum dilatatá; vertice truncato, depresso; spirá occultá; columellá obtusè uniplicatá.*

Shell elongato-cylindrical, convolute, upon a nearly horizontal axis, covered with fine, transverse, regular striæ; aperture linear, dilated at the base, with an obtuse fold upon the columella; vertex depressed, concave, with a hidden spire.

Axis, ⅝ of an inch; *diameter,* ⅔ its length.

Locality.   Cor. Crag, Sutton.

Red Crag, Sutton.   Recent, British Seas.

This species is very abundant at one locality in the Coralline beds, but is very rare as a Red Crag shell. The volution being nearly horizontal, the outer lip is carried above the spire, and there is a little inflection of the shell at that part which covers what would otherwise be a perforation; a small hole may be occasionally seen, but it is generally closed.

Among my Crag fossils, as well as among my recent specimens, are individuals both smooth and striated, but the condition of the latter is, I believe, dependent upon the state of preservation of the exterior. In a recent state it is covered by an epidermis, and then the striæ are well seen. In the description by the above-quoted authors,

* Since the above was written I have seen the recent shell from the Mediterranean, which is rather more cylindrical, less regularly fusiform, and the pointed portion of the upper lip less acuminated, with a few faint striæ upon the base; such differences are probably only the result of locality, and cannot be considered specific.

this shell is spoken of as perfectly smooth; but their examination must have been of ill-preserved individuals, the striæ being very minute and little impressed, they are easily obliterated.

A specimen from the Mammaliferous Crag, near Norwich (fig. 1 *b*), was found by Mr. Wigham; it has a diameter greater than three fifths of its axis, with a curved or semicircular outer lip, and is, I imagine, only a variety of this species.

5. BULLA CONCINNA. *S. Wood.* Tab. XXI, fig. 6, *a—c.*

BULLA CONCINNA. *S. Wood.* Illust. in Mag. Nat. Hist. 1839. p. 463, pl. 7, fig. 7.

— *S. Wood.* Catalogue 1842.

B. Testá minutá, subcylindraceá, tenui, fragili; transversim tenuissimè, striatá, et longitudinaliter decussatá; vertice umbilicato; spirá inclusá; columellá obsoletè uniplicatá.

Shell small, subcylindrical, thin, and fragile, finely striated transversely, and decussated by the lines of growth; vertex umbilicated; spire hidden; inner lip slightly replicate, with an obsolete fold upon the columella.

Axis, $\frac{1}{8}$ of an inch; *diameter,* $\frac{3}{5}$ of its length.

Locality. Cor. Crag, Sutton.

Unfortunately, since the publication of the description of this species at the above references, I have not found a specimen larger or more perfect than what I then possessed. It is very probable that this may be the young of *Bulla clathrata,* Defrance, *B. Tarbelliana,* Grateloup, Bord. foss. p. 58, f. 29-30. The principal difference appears to be in the size, although, from the figures, our shell is rather less cylindrical, with an obtuse fold upon the columella, which is most visible in broken specimens. I have left it with my own name for the present, until some better specimens are found to confirm either the one or the other.

6. BULLA TRUNCATA. *Adams.* Tab. XXI, fig. 3, *a—c.*

BULLA TRUNCATA. *Adams.* Linn. Trans. V, t. 1, fig. 1-2, 1797.

— — *Mont.* Test. Brit. p. 223, t. 7, fig. 5, 1803.

— RETUSA. *Maton* and *Rack.* Linn. Trans. vol. viii, 1807.

VOLVARIA RETUSA. *Brown,* Illust. Brit. Conch. pl. 38, fig. 17-18, 1827.

BULLA TRUNCATA. *Flem.* Brit. Animals, p. 293, 1828.

— SEMISULCATA. *Phil.* En. Moll. Sic. vol. i, p. 123, t. 7, fig. 9, 1836.

— SUBTRUNCATA. *S. Wood.* Illust. Mag. Nat. Hist. pl. 7, fig. 9, 1839.

— TRUNCATA. *Phil.* En. Moll. Sic. vol. ii, p. 96, 1844.

— — *S. Wood.* Catalogue 1842.

CYLICHNA TRUNCATA. *Lovén.* Ind. Moll. Scand. p. 10, 1846.

B. Testá pusillá, cylindricá, lævigatá; basi vix dilatatá; aperturá supernè angustè lineari; infernè paululum latiore; vertice truncato, vel latè umbilicato; spirá distinctá, plicá ad basim columellæ parum manifestá.

Shell minute, cylindrical, and smooth, with a little contraction in the middle, or rather dilated at the base; aperture linear, a little expanded below; vertex truncated, slightly depressed, with an obtuse almost invisible fold upon the base of the columella.

Axis, $\frac{1}{7}$ of an inch; *diameter*, $\frac{1}{16}$.

Locality. Cor. Crag, Sutton.   Recent, British, Mediterranean, and Arctic Seas.

This is rather scarce as a Coralline Crag fossil. The longitudinal lines of growth, so distinctly visible on the upper part of the recent shell, are not to be seen in my specimens, which are probably worn smooth. It differs from the young of *B. cylindracea* in having a distinct spire, in being smooth, transversely, and in its form, which is rather dilated below, like *B. Regulbiensis*, but it is more elongated than that shell, and appears to be intermediate between the two.

7. BULLA REGULBIENSIS. *Adams.* Tab. XXI, fig. 4, *a—c.*

BULLA REGULBIENSIS.   *Adams.* Micros. p. 640, t. 14, fig. 28, 1798.
—   —   *Turt.* Edit. Linn. vol. iv, p. 351, 1806.
—   OBTUSA.   *Mont.* Test. Brit. p. 223, pl. 7, fig. 3, 1803.
—   —   *Flem.* Brit. An. p. 293, 1828.
VOLVARIA REGULBIENSIS.   *Brown.* Illust. Brit. Conch. pl. 38, fig. 12, 1827.
RETUSA OBTUSA.   -   -   -   -   pl. 38, fig. 5-6.
—   PLICATA.   -   -   -   pl. 38, fig. 1-2.
—   DISCORS.   -   -   -   -   pl. 38, fig. 3-4.
BULLA MINUTA.   *Woodward.* Geol. of Norf. t. 3, fig. 1833.
—   OBTUSA.   *S. Wood.* Illust. in Mag. Nat. Hist. pl. 7, fig. 10, 1839.
—   —   *S. Wood.* Catalogue 1842.

B. Testá minutá, subcylindricá; spirá distinctá, obtusá, aliquando elevatá; anfractibus quatuor vel quinque, in mediis coarctatis; suturis profundis; aperturá ad basim dilatatá; labro arcuato.

Shell small, subcylindrical, with a distinct, sometimes slightly elevated, and obtuse spire; volutions four or five, occasionally contracted in the middle; suture deep; outer lip sharp and arched; aperture dilated below; inner lip spreading, and without a fold upon the columella.

Axis, $\frac{1}{8}$ of an inch.

Locality. Mam. Crag, Bramerton.   Recent, British Seas.

This species is, I believe, rare. It is, I presume, an Estuary species, as I have found it recent, out of the reach of salt water, at low tides. Like most of the shells from the Norwich Crag, the few specimens I have seen present considerable variation. Two individuals, sent me by Mr. Wigham for description, resemble the recent shell more than my own specimens, and indeed are rather more slender, while my own are less so. Lines of growth are distinctly visible, and in my recent shell some obscure transverse striæ may be seen upon the lower half. Fig. 4 *b* is from Mr. Wigham's specimen.

8. BULLA LAJONKAIREANA. *Bast.* Tab. XXI, fig. 5, *a—c.*

    BULLINA LAJONKAIREANA. *Bast.* Desc. Geol. du Bassin. Tert. du S. O. de la France, pl. 1, fig. 25, 1825.

    —   *Grat.* Not. sur la Fam. des Bull. p. 64, pl. 3, fig. 45-46, 1838.

    BULLA SPIRATA. *Dubois,* de Montp. Foss. de Volhyn. pl. 1, fig. 11-12, 1831.

    —   TEREBELLATA.   -   -   -   -   pl. 1, fig. 8-9.

    —   CLANDESTINA   -   -   -   -   pl. 1, fig. 19-21.

    —   MAMMILLATA. *Phil.* En. Moll. Sic. vol. 1, p. 122, t. 7, fig. 20, 1836.

    —   LAJONKAIREANA. *Desh.* 2d edit. Lam. Hist. des An. s. Vert. tom. vii, p. 678, 1836.

    —   —   *Dujard.* Mém. Soc. Géol. de France, tom. ii, pt. 2, p. 237, 1837.

    —   OLIVULA. *S. Wood.* Illust. Mag. Nat. Hist. p. 365, pl. 7, fig. 11, 1839.

    —   LAJONKAIREANA. *S. Wood.* Catalogue 1842.

B. Testá pusillá, cylindricá, lævigatá; spirá brevi, apice obtuso; suturis subcanaliculatis; aperturá supernè angustá, lineari, infernè dilatatá; plicá ad basim columellæ vix distinctá.

Shell minute, cylindrical, and smooth, with a conspicuous, slightly elevated spire, and mammilated apex; suture distinct, subcanaliculated; aperture linear, narrow at the upper part and dilated below, with an obscure fold upon the columella.

Axis, $\frac{1}{7}$ of an inch.

Locality. Cor. Crag, Sutton.   Recent, Mediterranean.

My cabinet contains about twenty specimens of this species, but unfortunately, all are in bad condition; there is, however, I think, no doubt about its identity. The lower part of our shell is rather narrower than the upper. It appears subject to a considerable variation in the amount of elevation of the spire, which would most probably produce a like variation in its cylindrical character. All my specimens are perfectly smooth, excepting the lines of growth.

9. BULLA NANA. *S. Wood.* Tab. XXI, fig. 13, *a—b.*

    BULLA NANA. *S. Wood.* Catalogue in An. and Mag. Nat. Hist. p. 461, pl. 5, fig. 1.

B. Testá pusillá, exiguá, lævigatá, hyaliná; spirá elevatá; apice obtusiusculo; anfractibus tribus convexiusculis; suturis profundis, subcanaliculatis; labro acuto, arcuato; aperturá ovatá; columellá leviter recedente, vix reflexá, umbilico parvo.

Shell very minute, thin, smooth, fragile, and semitransparent, with an elevated spire and rather obtuse apex; volutions slightly convex; suture deep, subcanaliculated; aperture ovate; outer lip sharp and curved, deeply sinuated at the upper part; inner lip slightly reflected, with a small but distinct umbilicus.

Axis, $\frac{1}{2}$ a line; *diameter,* $\frac{3}{4}$ the axis.

Locality. Cor. Crag, Sutton.

Of this delicate and pretty little shell I have found about half a dozen specimens, one of which shows a peculiarly curved outer lip, sinuated or deeply cleft near the suture, somewhat in the form of *Stylifer astericola*.

It slightly resembles *B. hyalina*, Turt., but is much smaller, having the volutions more convex, and without the flattened and angulated form of the upper part of the volution of that species, nor is the apex so papilliform.

BULLÆA. *Lamarck.*

PHILINE. *Ascanius.* 1772 (ex *Lovén*).
LOBARIA. *Müll.* 1776.

Gen. Char. Shell ovato-globulous, thin, and fragile, subhyaline or semitransparent in a recent state; slightly convolute, with a large and' open aperture, and expanded outer lip; spire small, sometimes hidden by the elevation of the outer lip.

In this genus the mantle is said by M. Deshayes, (second edition Lamarck, Hist. Nat. des An. s. vert., vii, p. 664,) to be exceedingly large, covering the entire shell, and rendering it invisible, and that the animal does not adhere by any muscle of attachment. In some species, however, a deep muscular impression is made, and left near the outer edge, both in the lower and upper part of it, as if the animal had adhered very strongly to its shell.

1. BULLÆA QUADRATA. *S. Wood.* Tab. XXI, fig. 9, *a—c.*

> BULLÆA QUADRATA. *S. Wood.* Mag. Nat. Hist. p. 461, pl. 7, fig. 1, 1839.
> BULLA QUADRATA. *S. Wood.* Catalogue in An. and Mag. Nat. Hist. p. 460, 1842.
> PHILINE SCUTULUM. *Lovén.* Ind. Moll. Scand. p. 9, 1846.

B. Testá pusillá, apertá, subquadratá, tenui; vertice truncato, subumbilicato; anfractu laxè convoluto, supernè coarctato, striis divergentibus, cateniformibus ornato; margine integro, infernè quadrato; labio reflexo, sine umbilico; aperturá patentissimá.

Shell minute, thin, and fragile, of a subquadrate form, with a large and expanded outer lip; vertex truncated; apex slightly depressed; volution contracted at the upper part, ornamented with diverging, punctured, or chain-like striæ; inner lip expanded over the umbilicus; outer lip sharp; muscular impression deep and distinct.

Axis, $\frac{1}{7}$; *diameter,* $\frac{1}{8}$ of an inch.

Locality. Cor. Crag, Sutton. Recent, British Seas.

I have but one perfect specimen of this shell, which may possibly be the same species as the one recently found and described by M. Lovén, in his 'Synopsis of the Scandinavian Mollusca.' The state of preservation of the exterior of my fossil is not

such as to permit of correct description, but it does not appear to correspond with the
" *fasciis binis incrassatis*" of M. Lovén.   Mr. Alder informs me that this species has
been found in a living state upon the coast of Northumberland.   A slight contraction
in the upper part of the volution takes from the otherwise nearly quadrate contour of
the shell.

2. Bullæa sculpta.   *S. Wood.* Tab. XXI, fig. 10, *a—c.*
     Bulla catenata.   *S. Wood.* Illust. in Mag. Nat. Hist. p. 460, pl. 7, fig. 2, 1839.
     Bullæa sculpta.   *S. Wood.* Catalogue 1842.

    *B. Testá pusillá, ovatá, tenui, fragili ; anfractibus supernè coarctatis, densissimè striatis;
striis divergentibus, sculptis, cateniformibus ; spirá depressá, vix distinctá ; aperturá
patente, labio acuto prominente, umbilico parvo.*

Shell small, ovate, thin, and fragile ; volutions slightly contracted round the upper
part ; spire depressed, scarcely visible ; covered externally with diverging chain-like
striæ ; outer lip curved ; inner lip sharp and prominent, with a small umbilicus ; muscular impression slightly visible.

Axis, $\frac{1}{6}$; *diameter,* $\frac{1}{8}$ of an inch.

Locality.   Cor. Crag, Sutton.

Only two or three specimens of this elegant shell have as yet come into my possession, and these appear somewhat to resemble *B. catena*, Montague ; but a specimen
of the latter, obligingly sent to me by Mr. Alder, differs in the following characters :
—it has a wider aperture, is less convolute, more quadrate, or more expanded, in the
lower part of the outer lip, and wants the inflected and prominent portion of the inner
lip ; a large series of both species might, perhaps, show a greater resemblance, but
there is sufficient difference in the specimens I possess to justify their being considered
as distinct.   *Bulla punctata*, Adams, Lin. Trans. v. t. 1, f. 6—8, is stated by Dr. Turton,
Mag. Nat. Hist. 1834, p. 353, to differ from *B. catena* in having distinctly-impressed
dots upon its surface, instead of oval, raised, chain-like striæ, which form the lines in
B. catena, and the animal is said to be destitute of a gizzard.   Our shell is covered
with regular chain-like striæ, similar to those upon *B. catena,* and was, no doubt,
white and semitransparent in its recent state.

3. BULLÆA SCABRA. *Müll.* Tab. XXI, fig. 12, *a—c.*
  LOBARIA SCABRA. *Müll.* Zool. Dan. t. 7, fig. 1, 1776.
  SCAPHANDER CATENATA. *Leach.* Moll. p. 55, 1820.
  BULLA PECTINATA. *Dillwyn.* Descrip. Catal. 1817.
  — *W. Wood.* Ind. Test. 1828.
  BULLINA GRANULOSA. *Sars.* Beskrivelser og Iagttagelser, p. 75, pl. 14, fig. 36, 1835.
  BULLA ANGUSTATA. *Phil.* En. Moll. Sic. p. 121, t. 7, fig. 17, 1836.
  BULLÆA PUNCTATA. - - vol. ii, p. 95, 1844.
  BULLA DILATATA. *S. Wood.* Illust. in Mag. Nat. Hist. p. 462, pl. 7, fig. 3, 1839.
  — CATENATA. *S. Wood.* Catalogue 1842.
  — LINEOLATA (?). *Gould.* Invert. Massachus. p. 169, fig. 99, 1841.
  PHILINE SCABRA. *Lovén.* Ind. Moll. Scand. p. 9, 1846.
  BULLA. Encyclopédie Méthodique, pl. 360, fig. 3.

B. Testá angustá, subcylindraceá, tenui fragili, infernè dilatatá ; densissimè striatá, striis divergentibus, cateniformibus ; spirá distinctá, convexá, obtusá ; labro acuto, sine umbilico.

Shell subcylindrical, thin, and fragile ; volutions loosely enrolled ; aperture elongate and gaping, expanded at the lower part, externally striated, with impressed chain-like diverging striæ ; outer lip sharp, a little curved ; inner lip slightly reflected, and without an umbilicus ; muscular impression distinct and deep.

Axis, ½ an inch.

Locality. Cor. Crag, Sutton.   Recent, Scottish Coast.

This is a very rare and an extremely fragile shell, and is no doubt identical with the northern British species, which, in the recent state, is nearly transparent ; that character is, of course, lost in the fossil. The upper part of the volution is rather rounded, causing a deep suture, or subcanal, around the spire, which is low and convex. The beautiful chain-like striæ encircle the upper part of the shell horizontally, but diverge towards the base, running parallel to the inner lip ; they are formed of small oval depressions, like the links of a chain, visible from within, and producing, at the base of recent individuals, a fimbriated edge. A slight contraction above the middle of the volution is visible in some specimens, but not in all. The aperture is expanded at the lower part, and a muscular impression is visible within the outer lip. This impression is separated in the centre, and deeply impressed both above and below, as if the animal adhered firmly to the shell ; the outer lip is rounded at the upper part, and the spire is always visible. The left lip is slightly reflected, forming a ridge, but there is no umbilicus. M. Philippi, in his second volume, has changed the name of *angustata* to that of *punctata,* Adams, which, I believe, is a very different shell.

4. BULLÆA VENTROSA.　*S. Wood.* Tab. XXI, fig. 11, *a—c.*

BULLA VENTROSA.　*S. Wood.* Illust. in Mag. Nat. Hist. p. 462. pl. 7, fig. 5, 1839.

　　　—　　　*S. Wood.* Catalogue 1842.

B. Testá pusillá, crassá, ventrosá, ovato-globosá; spirá occultá; anfractibus paucis, obliquis, laxè convolutis, supernè vix coarctatis, striatis; striis divergentibus, excavato-punctatis; aperturá ovali, infernè patente, dilatatá; labio prominente, pone umbilicato.

Shell small, strong, ventricose, ovato-globose; spire concealed by the elevation of the outer lip; volutions few, slightly enrolled, covered externally, with numerous regular, punctured, and diverging striæ: body of the whorl slightly contracted on the upper half; aperture ovate, somewhat contracted above, expanded below; inner lip prominent, slightly reflected, with a small umbilicus behind it.

Axis, $\frac{1}{7}$ of an inch; *diameter,* $\frac{1}{8}$.

Locality.　Cor. Crag, Sutton.

I have obtained only a few specimens of this shell.　It somewhat resembles a minute *B. lignaria,* but it has an umbilicus, and is otherwise distinct.　My specimens appear to have punctured lines, and not chain-like striæ on the exterior; but they may, perhaps, be in an altered condition.　This may possibly be *Bulla zonata,* Turt. Mag. Nat. Hist. 1834, p. 352, but the crown of our shell can scarcely be said to be umbilicated; the outer lip is deeply cleft or sinuated, and disconnected up to the vertex; and the replication of the inner lip forms a distinct umbilicus.　There is a deep muscular impression running nearly parallel with the outer lip, which is regular and continuous the whole length of the shell.

ORD. *CYCLOBRANCHIATA.* CUV.

PATELLA.　*Linnæus.* 1758.

Gen. Char.　Shell conical, clypeiform, with an oblong or ovate base; apex or vertex subcentral, always more or less on the anterior portion; generally rayed, striated, or costated, with a crenulated margin, occasionally smooth; muscular impression, a ligulate band surrounding the interior, about half way up, open in front for the head of the animal.

Animals of this genus differ from the fissured and keyhole limpets in having the branchiæ on the outer edge under the margin, and have, in consequence, been placed in a different order.　When living, the edge of the shell is generally slightly raised from the rock, to admit the water to the branchiæ, and the blade of a knife can easily be passed under the margin of the shell, if undisturbed; but when alarmed, the water and air are speedily expelled, and the shell then becomes so firmly fixed by atmospheric pressure, as to be removed with great difficulty. Shells of this form have early made their appearance, and several have been figured from the Secondary formations.

1. PATELLA VULGATA. *Linn.* Tab. XX, fig. 8, *a—b*.

> PATELLA VULGATA. *Linn.* Syst. Nat. p. 1258, 1758.
> — VULGARIS. *Da Costa.* Hist. Nat. Test. Brit. p. 3, t. 1, fig. 1-2, 1778.
> — VULGATA. *Mont.* Test. Brit. p. 475, 1803.
> -- — *Brown.* Illust. Brit. Conch. pl. 37, fig. 15, 1827.
> — — *Flem.* Brit. An. p. 216, 1828.
> — — *Lam.* Hist. des An. s. Vert. 2d edit. tom. 7, p. 535.
> — — *S. Wood.* Catalogue 1842.
> — — *Lovén.* Ind. Moll. Scand. p. 26, 1846.

P. Testá ovato-conicá, depressá ; vertice eccentrico ; costis tenuibus, subangulatis, imbricatis ; margine subdentato.

Shell shield-formed, ovate, and depressedly conical, with an eccentric vertex, covered externally with low subangulated and subimbricated rays; margin slightly crenulated.

Longest diameter, $\frac{3}{8}$ of an inch.

Locality. Red Crag, Sutton. Recent, Britain.

This species, so exceedingly abundant at the present day upon our own coasts, appears to have been very scarce during the Red Crag period, as only two specimens have as yet come into my possession; these are small, and differ slightly from the general character of the recent shell, in having the rays exceedingly rough or slightly imbricated; there is, however, I think, no doubt of its identity.

CHITON. *Linn.* 1758.

Gen. Char. Animal in a recent state covered with eight testaceous valves, generally broader than long, more or less arched or angulated. These valves are fixed at their lateral edges, in the coriaceous skin which forms a rim around them ; collectively, they are of an ovate, or oblongo-ovate form, rounded at the extremities, one valve slightly or partly overlapping the succeeding one, enabling the animal to roll itself up into a ball ; each valve furnished with two projecting processes.

As the eight valves (called by M. de Koninck *cerami*) of each individual in this genus are held in position by the connecting integuments, they are in a fossil state, as might be supposed, almost always detached and separated, at least in such sandy formations as the Crag. The band which connects and surrounds these valves, in the living state, is of a coriaceous nature, sometimes beset with prominent spines, sometimes rugose like shagreen, and often nearly smooth This is a good character for specific distinction, but an auxiliary not to be obtained in Crag specimens ; the valves, however, when well preserved, may often be distinguished by their sculpture, as well as by their form. For the information of those collectors to whom it is not known I may observe, that the anterior or *primal* valve has no projecting processes or sustentacula, and the centre or umbo is situated at the posterior margin of that valve; whereas, in the *final* vale, or that which covers the posterior extremity of the animal, the projecting portion or umbo is sometimes in the centre; and the anterior edge has a projecting process or sustentaculum* on each side, upon which the adjoining valve rests; similar processes project on the anterior edge of each succeeding valve, and support the one adjoining up to the primal valve, which covers the head. By the sliding of these valves over each other a considerable movement is given to the animal, which is enabled to roll itself up after the manner of the Oniscus; the whole length of the valve is only exposed when the back of the animal is at its greatest tension. The form and size of these sustentacula vary in all the species I have examined, and, when well preserved, will materially assist in specific determination. Each of the central valves is diagonally divided into two areas by a line more or less distinct, passing from a projecting central point or umbo at the posterior part to the anterior lateral edge   One is called the dorsal and the other the lateral area, and the shell is increased by the addition of calcareous matter deposited at the anterior and lateral edges of each valve from the central posterior point or umbo, and not around them, the external and ornamental markings being superadded to the sustentacula or apophyses, as they are called by M. Koninck, and which are always buried within the mantle. The primal valve, or that which covers the head, is generally of a hemisperical form, and may be considered as an enlarged dorsal area, with radiating markings, while the posterior or final valve has the lateral areas prolonged or extended round the termination behind the umbo.

This genus was instituted by Linnæus. Lamarck separated those eight-valved Cyclobranchiate animals in which the valves were placed upon the back at a distance from each other, to which he gave the name Chitonellus; and Mr. Salter, in the 'Quarterly Journal of the Geol. Society,' Feb. 1846, has formed another division for some secondary fossil and elongated Chitons, under the name of Helminthochiton, or worm-shaped Chitons, in which the valves are longer than wide, but that character is also found in some recent species. These latter shells are found in the Silurian or Protozoic Rocks, and the genus is probably continued through all the intermediate periods.

Animals belonging to this order are said to have a double generative system, and the branchiæ are situated under the margin of the shell. They are generally found adhering to stones or upon rocky coasts, seldom ranging deeper than twenty-five fathoms, and some are found above low-water mark.

M. de Blainville published a monograph of this family in the ' Dict. des Sci. Nat.' xxxvi, with a new arrangement, some of the characters of separation depending upon

---

* These projecting processes are called by Mr. Gray plates of insertion.

the soft parts of the animal, while the colour of the valves was also considered a distinction.

In the 'Zool. Journ.' for 1829, the Rev. Lansdown Guilding divided the Fam. *Chitonida* (Gray) into five genera, viz., *Chiton, Acanthopleura, Phakellopleura, Chito-nellus*, and *Cryptoconchus*.

As three of these are distinguished by the ornamental markings and appendages upon the coriaceous margin or zone which surrounds the shell, and the remaining two depend upon the position of the cerames or scutæ on the back of the animal, they possess an artificial character of no great value in the determination of fossils. Mr. Gray, in the Proc. of the Zool. Soc., 1847-48, has given a new arrangement of this family, and divided it into twenty genera.

1. CHITON FASCICULARIS (?). *Linn.* Tab. XX, fig. 9, *a—b.*

CHITON FASCICULARIS.	*Linn.* Syst. Nat. p. 1106, 1766.
—	*Chemn.* Conch. x, t. 173, fig. 1688, 1780.
—	*G. Sow. jun.* Conch. Illust. fig. 87-87*a*, 1841.
—	*Phil.* Enum. Moll. Sic. vol. i, p. 108, t. 7, fig. 2, 1836.
—	*S. Wood.* Catalogue 1842.
—	*Morris.* Catal. of Brit. Fossils, p. 142, 1843.

Ch. Testá subcarinatá, angulatá; valvarum carinis longitudinaliter striatis; lateribus subplanis, compressis, granulatis.

Shell subcarinate, angular; the central part or keel longitudinally striated; sides rather flat, covered with depressed granulations.

Dimensions (?)

Locality. Cor. Crag, Sutton. Recent, Britain.

I have but a few valves of this species, and those are not in good preservation. The sculpture and ornaments appear to correspond with the well-known recent British species, but the valves are more elevated and angular, and have a sharper back than any of my recent specimens. The sides of the fossil valves diverge at an angle of less than 90°, and the length is nearly equal to the breadth of one of the sides. The sustentacula or apophyses in all my specimens are unfortunately imperfect. There is generally in the recent specimens a considerable flatness in the anterior part of the dorsal area produced by the action of the posterior projecting umbo of the antecedent valve. The granules upon the lateral areas are large, and more or less elongato-oval, increasing in size from the young shell or umbo to the sides, and appear placed in irregular quincunx order. A curved, diverging linear arrangement may be observed in some specimens, and these granules are quite flat on the top, like those of *Ch. fascicularis*, and are larger than those upon *Ch. crinitus*, and the valves are elevated. The Crag shell has characters that appear to unite the two. Better specimens than those I possess are necessary to determine the true character. The dorsal area has what appear to be punctured striæ, like those upon the recent *C. fascicularis*.

24

2. CHITON STRIGILLATUS. *S. Wood.* Tab. XX, fig. 10, *a—b.*
      CHITON STRIGILLATUS. *S. Wood.* Catalogue 1842.

C. Testá (?), *ceramis dorsalibus angulatis, subcarinatis; areá dorsali magná, longitudinaliter et irregulariter striatá; striis rugosè granulatis; areis lateralibus parvis et granulatis; ceramo ultimo magno, umbone subcentrali; apophysibus magnis convexis.*

Shell (?), central valves angular and subcarinate, with a large dorsal area longitudinally striated; striæ rather irregular, diverging, and roughly granulate; lateral area small and granular; posterior or final valve large, with a large dorsal area, and nearly central umbo; terminal area of this valve granular, not striated.

Angular divergence of dorsal valve, 95°.

Locality. Cor. Crag, Sutton.

The valve figured at fig. 10 *a* is probably the penultimate one; it is long in proportion to its breadth, measuring about three fourths of the width of one side, that is, from the dorsal angle to the lateral edge. It is narrower towards the sides, with large projecting apophyses, convex or rather angular outwards. The dorsal area of this valve is covered with linearly arranged granular striæ, which diverge and curve, and increase in number, or present an intermediate one about midway. The granules in the terminal area of the final valve (fig. 10 *b*) appear to be concentrically arranged, though not in perfect regularity. In this valve the umbo is nearly as far from the anterior margin as it is from the extreme and terminal edge, giving it a subcircular form, the longer axis being in a lateral direction. It appears, in the form of the valves, to be somewhat like *Ch. lævis,* but it is differently ornamented, and I know of no recent species, with a final valve, so circular. The granular striæ resemble those of *Ch. cinereus,* but the valves in that species are much broader in proportion to the length. The granules appear to be hollow, as an opening is left in them where the surface is worn.

Not being able to identify this shell with any species I was acquainted with, a new name was provisionally given to it in my Catalogue, and this name has been retained for the present, and such a description given as the imperfect materials will admit of, which may perhaps, at some future time, assist a conchologist better acquainted with this genus than I am in the correct determination of the species.

3. CHITON RISSOI (?). *Payr.* Tab. XX, fig. 11, *a—c.*
      CHITON RISSOI. *Payr.* Cat. des Annelid. et des Moll. de l'ile de Corse, p. 87, pl. 3, fig. 4-5, 1826.
      — ARCUARIUS. *S. Wood.* Catalogue 1842.

C. Testá (?), *ceramis dorsalibus, brevibus, convexis, fornicatis, non carinatis; areá dorsali longitudinaliter striatá; striis minutè granulatis; areis lateralibus transversim striatis, striis radiantibus, granulatis; apophysibus subangulatis et prominentibus.*

Shell (?), with arched valves, and an indistinct, almost invisible, carina, covered with granular striæ; the dorsal area has the striæ longitudinally or linearly arranged, and on the lateral areas the striæ are transverse or diverging; apophyses subangular, projecting, and sublateral.

Altitude of arch, ½ the span.

Locality.   Cor. Crag, Sutton.   ? Mediterranean.

Single valves of this species are abundant.   It was considered in my Catalogue as a distinct species; I have now united it with the above Mediterranean species, from its general resemblance, although with considerable doubt.   The lateral areas have the granules in transverse or radiating striæ, not quite so regular and distinct as those upon the dorsal area; but the primal as well as the final valves have the different areas striated in a radiating manner.   The projecting processes are near the sides, and are of an obtusely-angular form, somewhat resembling the blade of a lancet used for bleeding cattle.   On the final valve they do not project so much, and extend further along the margin inwards.   Fig. 11 *a* is the primal valve; 11 *b*, one of the central valves; 11 *c*, the terminal or final valve.

There is every reason to believe that two more species of the genus Chiton existed during the Coralline Crag period, valves of which are in my cabinet.   They are not, however, in sufficiently good preservation for description, and their characters cannot be accurately determined.

Ord. *CIRRIBRANCHIATA.*

DENTALIUM.*   *Linn.* 1740.
TUBULUS *Scilla.*   (ex *Hermansen.*)
SOLEN (spec.)   *Rumph.*   Id.

Gen. Char. Shell tubular, symmetrical, elongato-conical, or subcylindrical, generally smooth, sometimes annulated, often costated, slightly curved, open at both ends, smaller at the posterior extremity, which is sometimes entire, sometimes with a medial and dorsal cleft, occasionally with two lateral indentations.

Animals of this genus are now determined to be true Mollusca, though belonging to a distinct order, having the branchiæ near the upper part of the neck, in the form of two bundles of fine, soft, flexible, tentacular filaments.   They are placed in a natural arrangement near the Patellæ.   M. Deshayes, in a Memoir read before the Société d'Hist. Naturelle, 1825, gave the anatomical details of one species, and a monograph of the genus.   In his description he places the shell with its small end downwards, considering it the posterior extremity; and the convex portion of the shell, as corresponding with the dorsal surface of the animal; the larger end, therefore, is the anterior.

* Etym. *Dentale,* from *dens,* a tooth.

He has formed one section of this genus containing those species in which the posterior extremity has a medial dorsal cleft, the nature of which is as yet but imperfectly understood; but I imagine this character will prove insufficient for specific determination.

1. DENTALIUM COSTATUM. *J. Sow.* Tab. XX, fig. *a—f.*

DENTALIUM COSTATUM. *J. Sow.* Min. Conch. t. 70, fig. 8, 1814.
— — *S. Wood.* Catalogue 1842.
— — *Morris.* Catalogue of Brit. Foss. p. 143, 1843.
— — *Nyst.* Coq. foss. de Belg. p. 344, pl. 35, fig. 2, 1844.
— DENTALIS (?). *Desh.* Monog. de Dent. p. 33, pl. 2, fig. 9-10, 1825.
— — *Gould.* Inv. of Massach. p. 156, fig. 1, 1841.
— STRIATUM. *Mont.* Test. Brit. p. 495, 1803.
Not. DENTALIUM FOSSILE. *Desh.* Monog. de Dent. p. 35, pl. 3, fig. 12.

D. Testá tereti, eleganter arcuatá, subulatá; longitudinaliter costatá, costis 10—18, *obtusis, convexis, eminentioribus; striis transversis minimis; extremitate posticá, subprofunde fissá; juventute acuminatá.*

Shell tubular, elegantly arched, and tapering; longitudinally costated; costæ obtuse, convex, and prominent, varying in number from ten to eighteen; finely striated transversely; anterior margin thin and sharp; posterior extremity with a dorsal cleft, acuminated and open, but not cleft, in the young state.

Length, $1\frac{7}{8}$ of an inch; *diameter*, nearly 2 lines.

Locality. Cor. Crag, Sutton.

Red Crag, Sutton.   Recent, Mediterranean.

Fragments or imperfect specimens of this species are abundant in the Coralline Crag, and they present a great variety in the number, magnitude, and arrangement of the costæ, which, in some specimens, are broader than the spaces between them, while in others they are narrower. They are always obtuse and rounded, often irregular in distance, and are continued from one extremity to the other, with sometimes a narrower intermediate line. In perfect specimens there is a medial dorsal cleft to the depth of a line; and in one individual the posterior extremity has not only a slit, but the terminal aperture is partially covered with a convex epiphragm, that has a cleft across it (fig. 1*f*), like the posterior termination of *D. fissura*, Sowerby's Genera. In small specimens, of what I presume to be the young of this species, the posterior termination is entire, *without* the cleft (fig. 1 *e*); it is evident, therefore, that the animal has the power of absorbing or reducing a portion of the younger shell, and of forming the cleft after it has attained a certain size. The anterior margin is thin and sharp, slightly oblique, projecting a little on the ventral side, with the lines of growth at all times visible, giving the appearance of having transverse striæ. A recent specimen of *D. dentale*, from the Mediterranean, given me by my friend Professor E. Forbes, corresponds with our shell in its external characters, but it is less in size, and has not the posterior cleft.

As that, I believe, is not a permanent character, I am much inclined to think the Crag shell is the same species. A fragment (fig. 1 *a*) is one third of an inch in diameter, with sixteen sharp angular costæ, and when perfect, must have measured three or three inches and a half in length. I imagine it to be only an altered form of the same species.

2. DENTALIUM ENTALE. Tab. XX, fig. 2, *a—b.*
    DENTALIUM ENTALIS. *Linn.* Syst. Nat. p. 1263, 1766.
    —   —   *Mont.* Test. Brit. p. 494, 1803.
    —   —   *Broc.* Conch. Subapenn. p. 263, No. 8, 1815.
    —   —   *Desh.* Monog. de Dent. p. 39, pl. 1, fig. 7, and pl. 2, fig. 2, 1825.
    —   LABIATUM.   *Brown.* Illust. Brit. Conch. pl. 1, fig. 4, 1827.
    —   ENTALIS.   -   -   .   pl. 1, fig. 7.
    —   —   *Dujard.* Mém. Soc. Géol. de France, tom. ii, pt. 2, p. 273, 1837.
    —   —   *Nyst.* Coq. foss. de Belg. p. 345, pl. 35, fig. 3, 1844.
    —   —   *Phil.* En. Moll. Sic. vol. ii, p. 206, 1844.
    —   —   *Thorpe.* Brit. Mar. Conch. p. 1, 1844.
    —   —   *Lovén.* Ind. Moll. Scand. p. 28, 1846.
    —   TARENTINUM.   *Lamarck.* Hist. des An. s. Vert. v. p. 345, 1818.

D. Testá tereti, leviter arcuatá, lævigatá, politá, crassiusculá.

Shell tubular, subcylindrical, slightly curved, smooth, glossy, thick, and strong, with distinct lines of growth; anterior margin acute.

Length,   (?)
Locality. Mam. Crag, Bridlington.   Recent, Britain.

One specimen from the above locality, among the fossils sent by Mr. Bean, is all that I have seen of this species. The posterior portion is broken off, and only about two fifths of an inch of the shell remaining, which, however, so fully corresponds with the recent species, that I think there is no doubt of its identity. This shell is quoted by M. Deshayes as fossil from Bordeaux, and also from the Eocene formations of France. The specimen figured by J. Sowerby, in 'Min. Conch.' t. 70, f. 3, is a doubtful identification. I have examined specimens from Barton and from Bracklesham, in the extensive collection of Mr. Edwards, and that gentleman agrees with me in the opinion that it is not a British Eocene fossil, at least, that we have not as yet seen anything that can be considered as *D. entale*, and that the figure in 'Min. Conch.,' as well as the one by Brander (fig. 9), are representations of specimens, the surfaces of which have been eroded, or, at least, have a less extended portion of the striæ, the striæ also being different from those upon *D. entale.* Fig. 2 *a* is a representation of what appears to be *D. tarentinum*, Lamarck, which M. Deshayes thinks is only a variety of the former; and as he has had the opportunity of examining the original specimens, upon which Lamarck founded his species *tarentinum*, I have given it upon his authority. The specimens figured are from the cabinet of Mr. Bean.

3. Dentalium bifissum. *S. Wood.* Tab. XX, fig. 3, *a—b.*
　　Ditrupa polita. *S. Wood.* Catalogue in Ann. and Mag. of Nat. Hist. 1842, p. 459, pl. 5, fig. 14.

D. Testá tereti, leviter arcuatá, subulatá, lævigatá, glabrá, politá; antice simplici, margine acutá ; postice bifidá in utroque latere fissá.

Shell tubular, tapering, and elegantly curved, smooth, and glossy; anterior extremity simple, not contracted ; margin acute, posterior termination bilaterally cleft, with the dorsal portion of the margin projecting and rounded, ventral portion short and truncate.

Length, ½ an inch.

Locality. Cor. Crag, Sutton.

Small and imperfect specimens of this shell are abundant in the Coralline Crag. In my Catalogue it was considered as a *Pteropod,* in consequence of my longest and largest specimen appearing to have a slight contraction at the anterior extremity, which, upon further examination, I believe to have arisen from a portion of the exterior having peeled off at that part, and as there is no one specimen in my collection that can be said to have a contraction at the anterior or larger extremity, it may, perhaps, be more correct to place it in this genus. This species, like *D. coarctatum,* Lam., has the posterior opening laterally cleft, but not very deeply. The portion on the dorsal side of the fissures projects considerably, and is rounded ; on the ventral side the edge is short and truncated. It differs from *D. coarctatum* in having this projecting terminal portion rather longer, and in the anterior part of the shell not being contracted. In this shell also the lines of growth are distinctly visible, and rather oblique to the axis of the shell ; in the Grignon specimens of *D. coarctatum* these lines are not visible under a common lens. The diameter of the posterior opening is very variable, though generally less than half that of the anterior.

I have no specimens without this peculiar termination, except where evidently broken off. I am, nevertheless, inclined to think that these clefts are produced by the animal at a certain stage of its existence, and that this peculiar form is not the state of the infant shell, as the lines of growth may be seen to cross the fissures, but that, like the young of the preceding species, the posterior termination was small and circular.

CLASS—PTEROPODA.

Ord. *THECOSOMATA.*

CLEODORA.*   *Peron.* and *Lessueur.* 1810.
    CLIO (spec.) *Linn.*
    HYALEA (spec.) *Lam.*
    VAGINELLA. *Daudin.*
    CLIODORA. *Schweig.*
    CRESEIS ET CUVIERIA. *Rang.* 1828.
    BALANTIUM. *Children.* (fide *Gray.*)
    CRISIA. *Menke.* 1844.

Gen. Char. Shell thin, fragile, and transparent; inversely pyramidal, posterior extremity closed and acuminated; anterior open, dilated in the centre, with lateral compressed aliform appendages.

In the recent state these shells are of a hyaline and vitreous texture, extremely thin and fragile, clear and crystalline as the thinnest and most transparent glass. That beautiful character has disappeared in the fossil, and they can scarcely be recognised in their altered condition.

The known animals belonging to this genus are all marine, and frequenters of deep and open seas, where they swim by means of wing-like fleshy fins, and appear like aquatic butterflies. The same power of locomotion is possessed by some of the Bullæ, which, in some of the characters of the animals, they much resemble. Species of this genus are found fossil in the Sicilian beds.

1. CLEODORA INFUNDIBULUM. *S. Wood.* Tab. XXI, fig. 14, *a—b.*
    CLEODORA INFUNDIBULUM. *S. Wood.* Catalogue, vol. ix, p. 459, pl. 5, fig. 13, 1842.

Cl. Testá subulato-conicá, lævigatá, infundibuliformi; antice apertá, dilatatá; postice elongatá, acutá, extremitate compressá, utrinque carinatá.

Shell elongate, tapering, trumpet-shaped, smooth, and fragile; posterior part elongated, and acuminated, with a compressed extremity; anterior open, dilated, sides

* Etym. *Cleodora*, a proper name. An. vero *Cliodora?* Hermansen.

sulcated; a sharp and angulated keel extends the whole length of the shell on each side.

Length, ¼ of an inch ; *transverse diameter,* one line.

Locality. Cor. Crag, Sutton.

This species is at present rare, most likely from its extreme fragility.   In all probability the posterior extremity was closed and pointed, but in each of my specimens it is open.   The sharp and sheath-like termination is not cylindrical, but compressed before and behind, leaving a carinated edge on each side, which is continued all the way up the shell.   There is a sulciform depression on both sides of the expanded part, before as well as behind ; those portions are not, however, of equal magnitude.   It was no doubt transparent in the recent state, although it looks now like an opaque shell, and is not very thin.

AUTHORITIES CONSULTED OR REFERRED TO.

ADAMS (J.) Specific Characters of some minute shells discovered on the coast of Pembrokeshire. (In Transactions of the Linnæan Society, vol. iii, 1797.)

ADANSON. Histoire Naturelle du Sénégal.—Coquillages. 4to, 1757.

AGASSIZ (L.) Etudes Critiques sur les Mollusques fossiles, 4to, 1842.

— Mémoire sur les Moules de Mollusques vivants et fossiles, 1839.

ALDER (J.) Descriptions of some new species of Rissoa and Odostomia. Annals and Magazine of Natural History, 1844.

BASTEROT. Description Géologique du Bassin Tertiare du S. O. de la France, 4to, plates, 1825. (Mém. de la Soc. d'Hist. Nat. de Par.)

BEAN. Notice of a Marine Deposit at Bridlington. Mag. Nat. Hist. vol. viii, 1839.

BIVONA (Bernardi.) Nouvi Generi et nouve Specie di Molluschi, 8vo, 1832.

BLAINVILLE (H. M. De.) Manuel de Malacologie et de Conchyliologie, 8vo, 1825.

BORN. Testacea Musei Cæs. Vindobonensis, fol. 1780.

BOWDICH (T. E.) Elements of Conchology, including the Fossil Genera and the Animals, 8vo, 1822.

BRARD. Histoire des Coquilles terrestres et fluviatiles qui vivent aux Environs de Paris, 8vo, 1815.

BRANDER. Fossilia Hantoniensia collecta, 4to, 1766. Reprint. W. Wood, 1829.

BROCCHI. Conchiologia fossile Subapennina, 4to, 1814.

BRONGNIART (Al.) Mémoire sur les Terrains du Sédiment supérieur calcareo-trappeen du Vincentin, 4to, 1823.

BRONN. Lethæa Geognostica, 2 vols. 8vo, atlas in 4to, 1837-38.

BROOKES (Sam.) An Introduction to the Study of Conchology. 1815.

BROWN (T.) Illustrations of the Conchology of Great Britain and Ireland, 4to, 1827.

— Descriptions of the Testacea of Ireland, in Memoirs of the Wernerian Society, vol. ii, pt. 2.

BRUGUIÈRE. Histoire Naturelle des Vers (Encyclopédie Méthodique), 1789-92.

BUCH (Leop. von.) Mémoire sur les Brachiopodes dans le 3e tome des Mém. de la Soc. Géol. de France, 1838.

BURROW (E. J.) Elements of Conchology according to the Linnæan System, 8vo, 1815.

CAILLIAUD (F.) Voyage à Meroé au Fleuve Blanc, fol. Paris, 1823.

CATLOW (Agnes), and Lovell REEVE. Conchological Nomenclator, 8vo, 1845.

CANTRAINE. Malacologie mediterranéene et littorale, ou description des Mollusques qui vivent dans la Mediterranée, 4to, 1840. (Inséré, dans les Mém. de l'Acad. Roy. de Brux. tom. xiii.)

CHEMNITZ et MARTINI. Systematische Konchylien Kabinet, 4to, 1780-95.

CLARK (W.)　Observations on the animals of some species of Bullæa, in Zool. Journ. 1827.
CROUCH (E. A.)　Illustrated Introduction to Lamarck's Conchology, 4to.
CUVIER (G.)　Le Règne Animal distribué après son Organisation, 1829.

DA COSTA (E. M.)　Historia Naturalis Testaceorum Britanniæ, 4to, 1778, 17 plates.
DALE (J.)　A Natural History of the Sea Coast and Country about Harwich, 4to, 1730.
DEFRANCE.　Tableau des Corps organisés, 8vo, 1824.
DEKAY (J. E.)　Natural History of New York, 11 vols. 4to.　Zoology.
DELLA CHIAJE.　Testacea utriusque Siciliæ, 4to, 1826.
DESHAYES.　Description des Coquilles fossiles des Environs de Paris, 2 vols. 4to, 1837.
——　Traité élémentaire de Conchyliologie, 8vo, 1839.
——　Description des Coquilles fossiles recueillies en Crimée, par M. de Verneuil, 4to, 1838 ; suivi d'Observations sur les fossiles de cette Péninsule, par M. DESHAYES (Mém. de la Soc. Géol. de France.)
——　Anatomie et Monographie du Genre Dentale, 4to, 1825.
Dictionnaire des Sciences Naturelles, 1816-29.
DILLWYN.　A descriptive Catalogue of Shells, 2 vols. 8vo, 1817.
——　Memoir on some Fossil Shells, in Philosophical Transactions 1823.
DONOVAN (Edw.)　Natural History of British Shells, 7 vols. 8vo, 1799-1803.
D'ORBIGNY (Alc.)　Palæontologie Française, 8vo, 1840-47.
——　Monographie d'un nouveau Genre de Mollusques Gasteropodes de la famille des Trochoïdes.　Gen. Scissurella. 4to, 1823.　(Mém. de la Société d'Histoire Naturelle de Paris.)
——　Voyage dans l'Amérique Méridionale, 1835-42, 4to, Paris.
DRAPARNAUD.　Histoire Naturelle des Mollusques terrestres et fluviatiles de la France, 4to, An. xiii, (1805.)
DUBOIS DE MONTPEREUX.　Conchyliologie fossile et aperçu géognostique des Formations du Plateau Wolhyni-Podolien.　Berlin, 4to, 1831.
DUJARDIN (F.)　Mémoire sur les Couches du Sol en Touraine, et description des Coquilles des faluns, 4to, 1837.　Dans le 2e tome des Mémoires de la Société Géologique de France.

Encyclopédie Méthodique.　Mollusques Testacés, parts 21 and 23, plates, An. vi.

FABRICIUS (O.)　Fauna Grœnlandica, 8vo, 1780.
FERUSSAC.　Histoire Naturelle générale et particulière des Mollusques terrestres et fluviatiles, fol. 1819-33.
FLEMING (J.)　Art. Conchology, in Edinburgh Encyclopædia, 1817.
——　Philosophy of Zoology, 8vo, 1822.
——　History of British Animals, 8vo, 1828.
FORBES (Edw.)　Malacologia Monensis.　A Catalogue of the Mollusca inhabiting the Isle of Man, and the neighbouring Sea.　Edinburgh, 8vo, 1838.
——　Geological Report upon the existing Fauna and Flora of the British Isles, 8vo, 1846.

GMELIN.　Linnæi Systema Naturæ, 8vo, edit. 13, 1788.
GOLDFUSS (G. A.)　Petrefacta Germaniæ, fol. 1827-44.
GOULD.　Report on the Invertebrata of Massachusetts, 8vo, 15 plates, 1841.
GRATELOUP.　Conchyliologie fossile du Bassin de l'Adour, ou description des Coquilles fossiles qui ont été trouvécs dans les terrains marins tertiares, aux environs de Bordeaux, 8vo, 1837. (Actes de la Soc. Linn. de Bord.)
——　Conchyliologie Fossile des Terrains Tertiaires du Bassin de l'Adour, 4to, tom i, Univalves et Atlas, 1847.

GRATELOUP. Notice sur la Famille des Bulléens, 8vo, 1838.   (Actes de la Soc. Linn. de Bord.)

GRAY (J. E.) Turton's Manual of Land and Freshwater Shells of Great Britain, new edition, thoroughly revised, 1844.

—   Description of Shells in Griffiths's edition of Cuvier, 1834.

—   Monograph of Cypræa, and Additions in Zoological Journal, 1829.

GRAY (M. E.) Figures of Molluscous Animals, 8vo, 1842.

HARTMAN VON HARTMANSRUTHE.   Syst. der Erd. und Flus. der Schweiz, 1821.

HARTMAN (J. W. D.) Erd und Süsswater-Gasteropoden, 4to, 1840-44.

HERMANSEN (A. N.) Indicis Genera Malacozoorum Primordia, 8vo, 1847.

HINDS (R. B.) Zoology of the Voyage of H. M. Ship Sulphur. Mollusca. 1844-5.

HISSINGER. Lethæa Suecica, 4to, 1837.

HUMPHREY. Museum Calonneanum, 1797.

JAY (J. C.) Catalogue of Shells, and descriptions of New and Rare Species, 4to, 1839.

JEFFREYS (J. G.) A Synopsis of the Pneumonobranchiate Mollusca of Great Britain.   (In Trans. Linn. Soc. vol. xvii, 1830.)

—   Additions to the British Testacea.   (In Annals and Mag. Nat. Hist. 1847.)

JOHNSTON (G.) Descriptions of some new Shells. (In Hist. of Berwickshire Natural History Club, 1834.)

Journal (Quarterly) of the Geological Society, 1845-47.

— Zoological, 1825-32.

KIENER. Spécies général et Iconographie des Coquilles vivantes, 8vo, 1835-47.

LAMARCK. Description des Coquilles fossiles des Environs de Paris, 4to, 1823.

—   Histoire Naturelle des Animaux sans Vertébres, 1ʳᵉ· édit. 7 vols. 8vo, 1815-22.

—   Id. 2ᵉ· édition, par MM. MILNE EDWARDS et G. P. DESHAYES, 8vo, 11 vols. 1835-45.

LASKEY. An Account of North British Testacea.   (Mem. of the Wern. Society, vol. i, 1811.)

LEA (J.) Contributions to Geology, 8vo, 1833.

—   Observations on the Genus Unio, 4to, 1832-42.

LEACH (W. E.) Synopsis of the British Mollusca, 8vo, 1820 (a small part only printed).

LESSON. Illustrations de Zoologie, 8vo, 1834-35.

LINNÉ. Systema Naturæ, 8vo, 12th edit. 1766.

LISTER (Martin.) Historiæ Conchyliorum, fol. 1685-88.

LOVÉN (S.) Index Molluscorum Scandinaviæ, 8vo, 1846.

LYELL (C.) Principles of Geology, 8vo, edit. 1846.

—   Elements of Geology, 2d edit. 2 vols. 12mo, 1841.

—   On the relative Ages of the Tertiary Deposits commonly called Crag. (Mag. Nat. Hist. 1839.)

MACGILLIVRAY. Molluscous Animals of Aberdeen, 18mo, 1843.

Magazine of Natural History, conducted by J. C. Loudon. 2d Series, contiuued by E. Charlesworth, 1829-41.

MARCEL (de Serres.) Géognosie des Terrains Tertiares du Midi de la France, 8vo, 1829.

MATON (W. G.,) M.D., and the Rev. J. RACKET. A descriptive Catalogue of the British Testacca. (In Trans. of Linn. Soc. vol. viii, 1807.)

MAWE (J.) The Linnæan System of Conchology, 8vo, 36 plates, 1823.

Mémoires de la Société Géologique de France, 1833-45.

MENARD DE LA GROYE. Mémoire sur un nouveau Genre de Coquille de la Famille des Solenoides. (An. du Mus. d'Histoire Naturelle, tom. ix, 1807.)

MICHAUD. Description de plusieurs nouvelles Espèces de Coquilles du Genre Rissoa, 8vo, 1832.

—   Complément de l'Histoire naturelle des Coquilles terrestres et fluviatiles de la France, 4to, 1831.

MICHELOTTI. Monographia del Genere Murex, 4to, 1841.

MONTAGUE. Testacea Britannica, 4to, 1803, and Supplement 1808.

MONTFORT (Denys.) Conchyliologie systématique, et Classification méthodique des Coquilles, 8vo, 1808-10.

MÖLLER (H. P. C.) Index Molluscorum Grœnlandiæ, 8vo, 1842.

MORRIS (J.) Catalogue of British Fossils, 8vo, 1843.

MÜLLER (O.) Zoologie Danicæ Prodromus, 1766.

—   Vermium terrestrium et fluviatilium historia, vol. i, 1773; vol. ii, 1774.

NYST (P. H.) Description des Coquilles et des Polypiers fossiles des Terrains Tertiares de la Belgique, 4to, 1843-46.

OLIVIER. Voyage dans l'Empire Ottoman, 8vo, 1801-7.

PARKINSON. An Introduction to the Study of Fossil Organic Remains, 8vo, 1822.

—   Organic Remains of a Former World, 3 vols. 4to, 1804-11.

PAYRAUDEAU (B. C.) Catalogue descriptif et méthodique des Annélides et des Mollusques de l'Ile de Corse, 8vo, 1826.

PENNANT (T.) British Zoology, 3d edit. 8vo, 1776.

PERON et LESSUEUR. Mémoires sur les Pteropodes. (Annales du Muséum de Paris, tom. xv, 1810.)

PETIVER (J.) Opera Historiam Naturalem Spectantia, or Gazophylacium, fol. 1764.

PFEIFFER. Naturgeschichte Deutscher Land und Süsswasser Mollusken, 4to, 1821-28.

—   Monographia Heliceorum viventium, 8vo, 1847.

PHILIPPI (R. A.) Enumeratio Molluscorum Siciliæ, 4to, vol. i, 1836.

—   Id. - - - 4to, vol. ii, 1844.

PICTET (F. J.) Traité élémentaire de Palæontologie, ou Histoire naturelle des Animaux fossiles, 4 vols. 8vo, 1844-5.

POLI. Testacea utriusque Siciliæ, 2 vols. fol. 1791-93.

QUOY et GAIMARD. Voyage de l'Astrolabe (Mollusques), 2 vols. 8vo, atlas fol. 95 plates, 1830.

RANG. Manuel de l'Histoire naturelle des Mollusques et de leurs Coquilles, 18mo, 6 plates, 1829.

REEVE (Lovell.) Conchologia Iconica, 11 parts, 4to, 1843-46.

Reports of the Meetings of the British Association for the Advancement of Science, 1831-47.

RISSOA. Histoire Naturelle des principales Productions de l'Europe Méridionale, 5 vols. 8vo, 1826.

RÖSSMASLER (E. A.) Iconographie der Lund und Süsswasser Mollusken, xii-liv. 8vo. Dresden and Leipz. 1835-44.

SARS (M.) Beskriv. og Jagttagelser, 4to, 1835.

SAVIGNY. Grand Ouvrage sur l'Egypte. Mollusques. 1826.

SCACCHI. Catalogus Conchyliorum Regni Neapolitani, 8vo, 1836.

SCHLOTTHEIM (Baron von.) Die Petrefactenkunde, 2 vols. 8vo. Gotha, 1820-22.

SCHUMACHER (C. P.) Essai d'un nouveau Système des Habitations des Vers testacés, 1 vol. 4to. Copenhagen, 1817.

SHEPPARD (Revett.) Descriptions of seven new British Land and Freshwater Shells, with observations upon many other species; including a List of such as have been found in the County of Suffolk. (Trans. Linn. Soc. vol. xiv, 1825.)

SMITH (J.) On the last Changes of the relative Levels of the Land and Sea in the British Isles. (Mem. Wern. Soc. vol. viii, 1838.)

—   On the Age of the Tertiary Beds of the Tagus. (Journ. Geol. Soc. Nov. 1847.)

SMITH (W.) Strata identified, 4to, 1816-19.

SOWERBY (G. B.)   Genera of Recent and Fossil Shells, 8vo, plates.
—   Malacological and Conchological Magazine, 8vo, 1839.
—   Catalogue of Shells in the Tankerville Collection, 8vo, 1825.
SOWERBY (G. B., jun.)   Conchological Manual, 2d edit. 8vo, 1843.
—   Conchological Illustrations, 8vo, 1836-41.
—   Thesaurus Conchyliorum, 1842-47.
SOWERBY (J. D. C.)   Mineral Conchology of Great Britain, 8vo, 1812-47.
SWAINSON (W.)   Treatise on Malacology, 12mo.   London, 1840.
—   Zoological Illustrations, First and Second Series, 8vo, 1820-33.

TENNANT (J.)   A Stratigraphical List of British Fossils, 18mo, 1847.
THOMPSON (W.)   Report upon the Invertebrata of Ireland, 1843.
THORPE (C.)   British Marine Conchology, 12mo, 1844.
Transactions of the Geological Society of London, 1826-41.
—   of the Zoological Society of London, 1833-44.
—   of the Linnæan Society of London, 1808-47.
TURTON (W.)   A Conchological Dictionary of the British Isles, 18mo, 1819.
—   Conchylia Insularum Britannicarum, 4to, 1822.
—   A General System of Nature, by Sir Charles Linné, 7 vols. 8vo, 1806.

VALENCIENNES.   Description de l'Animal de la Panopée Australe, et Recherches sur les Espèces du Genre
           Panopœa.   (Archives du Muséum de Paris, tom. i, 1839.)

WALKER (G.), and BOYS.   Testacea minuta rariora nuperrimè detecta, 4to, 1787.
WODARCH.   Introduction to the Study of Conchology, 4th edit. by J. MAWE, 12mo, 1832.
WOOD (W.)   Index Testaceologicus, 8vo, 1825.   Sup. 1828.
WOOD (S. V.)   Catalogue of Crag Shells, in Ann. and Mag. of Natural History, 1840-42.
—   Descriptions and Illustrations of Crag Shells.   (Magazine of Natural History, 1839.)
WOODWARD (J.)   An Attempt towards the Natural History of the Fossils of England, 8vo, 1729.
WOODWARD (S.)   British Organic Remains, 8vo, 1830.
—   Outline of the Geology of Norfolk, 8vo, 1833.

LOCALITIES at which Sections of the CRAG FORMATIONS have been visited, and Specimens obtained by the Author.

MAMMALIFEROUS CRAG.

Bramerton, near Norwich . . .	*Norfolk.*
Postwich ,, ,, . . .	,,
Thorpe ,, ,, . . .	,,
Thorpe, near Aldborough . . .	*Suffolk.*

RED CRAG.

Alderton	*Suffolk.*
Bawdsey	,,
Bealings	,,
Brightwell	,,
Brockstead	,,
Bromswell	,,
Butley	,,
Chillesford	,,
Chelmondiston	,,
Felixstow	,,
Foxhall	,,
Hollesley	,,
Holywells, near Ipswich . . .	,,
Kesgrave	,,

RED CRAG.

Levington	*Suffolk.*
Melton	,,
Newbourn	,,
Ramsholt	,,
Shottisham	,,
Sutton	,,
Tunstall	,,
Walton-on-the-Naze	*Essex.*
Woodbridge	*Suffolk.*

CORALLINE CRAG.

Aldborough	*Suffolk.*
Gedgrave	,,
Iken	,,
Ramsholt	,,
Sudbourn	,,
Tattingstone	,,
Orford	,,
Sutton	,,

ADDITIONAL LOCALITIES.

MAMMALIFEROUS CRAG.

Bridlington . . . . *Yorkshire.*
(On the authority of Messrs. Bean, Leckenby, and Charlesworth.)

Bulcham, near Southwold . . . *Suffolk.*
(On the authority of Capt. Alexander.)

RED CRAG.

Beaumont . . . . . . *Essex.*
(On the authority of John Brown, Esq.)

Harwich . . . . . . ,,
(As reported to have existed in the time of Dale, 1730.)

A SYNOPTICAL TABLE

OF

MOLLUSCA FROM THE CRAG.

— signifies not present.   × signifies present.

CLASS—GASTEROPODA.	Eocene.	Miocene.	Pliocene.	Pleistocene.	Recent.
ORD. PULMONATA.					
Helix hispida	—	—	—	×	×
" pulchella	—	—	×	—	×
" arbustorum (?)	—	—	—	×	×
" rysa	—	—	×	—	—
5. Cyclostoma elegans	—	×?	—	—	×
Succinea putris (?)	—	—	—	×	×
" oblonga	—	—	—	×	×
Limnæa palustris (?)	—	—	—	×	×
" peregra	—	—	—	×	×
10. " truncatula (?)	—	—	—	×	×
Planorbis complanatus	—	—	×	×	×
" spirorbis	—	—	—	×	×
" corneus	—	—	—	×	×
Conovulus pyramidalis	—	—	×	×	—
15. " myosotis (?)	—	—	×	×	×
ORD. PECTINIBRANCHIATA.					
SECT. a. SIPHONOSTOMATA.					
Ovula Leathesii	—	×	×	—	×
Cypræa avellana	—	×	×	—	—
" affinis	—	×	—	—	—
" Angliæ	—	—	×	—	—
20. " retusa	—	×	×	—	—
" Europæa	—	×	×	—	×
Erato lævis	—	×	×	—	×
" Maugeriæ	—	×	×	—	×
Voluta Lamberti	—	×	×	—	—
25. Mitra plicifera	—	×	×	—	—
Ringicula buccinea	—	×	×	—	—
" ventricosa	—	×	×	—	—
Columbella sulcata	—	—	×	—	—
Rostellaria plurimacosta	×?	—	×	—	—
30. Aporrhais pespelicani	—	×	×	—	×
Terebra inversa	—	×	×	—	—
" canalis	—	×	—	—	—
Cassidaria bicatenata	—	×	×	—	—
Nassa labiosa	—	×	×	—	—
35. " incrassata	—	×	×	—	×
" granulata	—	×	×	—	—
" propinqua	—	—	×	—	—
" elegans	—	—	×	—	—
" consociata	—	×	×	—	—
40. " Monensis	—	—	×	×	—
" conglobata	—	—	×	—	—
" prismatica	—	×	×	—	×
" reticosa	—	—	×	—	—
Buccinum Dalei	—	×	×	—	×
45. " undatum	—	×	×	×	×
Purpura lapillus	—	—	×	×	×
" tetragona	—	—	×	—	—
Murex erinaceus	—	—	—	×	×
" tortuosus	—	×	×	—	—
50. Triton heptagonum	—	×	—	—	—
Pyrula reticulata	—	×	—	—	×
Trophon antiquum	—	—	×	×	×
" elegans	—	—	×?	—	—
" gracile	—	×	×	×	×
55. " altum	—	—	×	—	—
" scalariforme	—	—	×	×	×
" costiferum	—	×	×	—	—

CLASS—GASTEROPODA.	Eocene.	Miocene.	Pliocene.	Pleistocene.	Recent.
Trophon alveolatum	—	×	×	—	—
" consociale	—	×	×	—	—
60. " imperspicuum	—	×	—	—	—
" muricatum	—	×	×	—	×
" gracilius	—	×	×	—	—
" paululum	—	×	—	—	—
Fusus intortus (?)	×	—	×	—	—
65. " porrectus (?)	×	—	×	—	—
Pleurotoma intorta	—	—	×	—	—
" turricula	—	—	×	—	—
" carinata	—	×	×	—	—
" semicolon (?)	×	×	×	—	—
70. " porrecta	×?	×	—	—	—
" nodulosa (?)	×	×	—	—	—
Clavatula linearis	—	×	×	—	×
" Philberti	—	×	—	—	×
" castanea (?)	—	×	—	—	×
75. " perpulchra	—	×	—	—	—
" costata (?)	—	×	×	—	×
" mitrula	—	×	×	—	—
" brachystoma	—	×	—	—	×
" nebula	—	—	×	—	×
80. " cancellata	—	×	×	—	×?
" concinnata	—	×	—	—	—
" lævigata (?)	—	×	×	—	×
" turricula	—	—	×	×	×
" Trevelliana	—	—	×	×	×
85. " Boothii	—	—	×	—	×
" plicifera	—	—	×	—	—
Cancellaria coronata	—	—	×	—	—
" mitræformis	—	×	×	—	—
" costellifera	—	×	×	×	×
90. " subangulosa	—	×	—	—	—
" læviuscula (?)	×	—	×	—	—
Trichotropis borealis	—	×	—	×	×
Cerithium tricinctum	—	—	×	×	—
" variculosum	—	—	×	—	—
95. " trilineatum	—	×	—	—	×
" tuberculare	—	×	—	—	×
" metaxa (?)	—	×	—	—	×
" cribrarium	—	×	—	—	—
" perpulchrum	—	×	—	—	—
100. " adversum	—	×	—	—	×
" granosum	—	×	×	—	—
SECT. β. HOLOSTOMATA.					
Turritella communis	—	—	×	×	×
" imbricataria	×	—	×	—	—
" incrassata	—	×	×	—	×
105. " planispira	—	×	—	—	—
" clathratula	—	—	—	×	×?
Pyramidella læviuscula	×?	×	—	—	—
Chemnitzia curvicostata	—	×	—	—	—
" rufa	—	×	—	—	×
110. " costaria	—	×	—	—	—
" nitidissima (?)	—	×	—	—	×
" elegantissima	—	×	—	—	×
" internodula	—	×	×	—	—
" filosa	—	×	—	—	×
115. " densecostata	—	×	—	—	×
" unica (?)	—	×	—	—	×
" varicula	—	×	—	—	—
" similis	—	×	—	—	×

SYNOPTICAL TABLE.

CLASS—GASTEROPODA.	Eocene.	Miocene.	Pliocene.	Pleistocene.	Recent.
Odostomia plicata	—	×	—	—	×
120. „ pupa	—	×	—	—	—
„ pellucida	—	×	—	—	×
„ similima (?)	—	×	—	—	×
Litiopa papillosa	—	×	—	—	—
Scalaria Groenlandica	—	—	×	×	×
125. „ varicosa	—	×	—	—	×
„ fimbriosa	—	×	—	—	—
„ hamulifera	—	×	—	—	—
„ frondosa	—	×	—	—	—
„ frondicula	—	×	—	—	—
130. „ foliacea	—	×	×	—	—
„ subulata	—	×	—	—	—
„ clathratula	—	×	—	—	×
„ Trevelyana	—	—	×	—	×
„ cancellata	—	×	—	—	—
135. „ obtusicostata	—	×	—	—	—
Eulima polita	—	×	×	—	×
„ subulata	—	×	—	—	×
„ glabella	—	×	—	—	—
Alvania ascaris	—	×	—	—	×
140. Rissoa striata	—	×	—	—	×
„ zetlandica	—	×	—	—	×
„ semicostata	—	—	—	×	×?
„ vitrea	—	×	—	—	×
„ punctura (?)	—	×	—	—	×
145. „ reticulata (?)	—	×	—	—	×
„ pulchella (?)	—	—	×	—	×
„ confinis	—	×	—	—	—
„ obsoleta	—	×	—	—	—
„ crassistriata	—	×	—	—	—
150. „ costulata	—	×	—	—	—
„ supracostata	—	×	—	—	—
Paludestrina subumbilicata	—	—	—	×	×
„ ulvæ	—	×?	—	—	×
„ ? terebellata	—	—	×	—	—
155. „ ? pendula	—	—	×	—	—
Paludina lenta	×	—	—	×	×
„ tentaculata	—	—	—	×	×
Valvata piscinalis	—	—	—	×	×
Vermetus intortus	—	×	×	—	×?
160. „ Bognoriensis (?)	×	—	×	—	—
Cæcum trachea	—	—	—	—	×
„ mammillatum	—	×	—	—	—
„ glabrum	—	×	—	—	×
„ incurvatum (?)	—	×	—	—	×
165. Littorina littorea	—	—	×	×	×
„ ? suboperta	—	—	×	—	—
Fossarus sulcatus	—	×	—	—	×
Lacuna reticulata	—	×	—	—	—
Turbo sphæroidea	—	×	—	—	—
170. Trochus crenularis	×	×	—	—	—
„ ziziphinus	—	×	×	—	×
„ conulus	—	×	—	—	×
„ formosus	—	×	×	—	×
„ papillosus (?)	—	—	×	—	×
175. „ subexcavatus	—	—	×	—	—
„ millegranus	—	×	—	—	×
„ multigranus	—	—	×	—	—
„ villicus	—	×	×	—	×
„ Montacuti	—	×	×	—	×
180. „ Adansoni	—	×	×	—	×
„ Kicksii	—	×	×	—	—
„ tumidus	—	—	×	×	×
„ cinerarius	—	—	×	—	×
„ cineroides	—	—	×	—	—
185. „ tricariniferus	—	×	—	—	×?
„ obconicus	—	×	—	—	×?
„ ditropis	—	×	—	—	—
Margarita elegantissima	—	—	—	×	—
„ maculata	—	×	—	—	—
190. „ trochoidea	—	×	—	—	—
Adeorbis striatus	—	×	—	—	—
„ supranitidus	—	×	—	—	—

CLASS—GASTEROPODA.	Eocene.	Miocene.	Pliocene.	Pleistocene.	Recent.
Adeorbis tricarinatus	—	×	×	—	—
„ subcarinatus	—	×	×	—	×
195. „ pulchralis	—	×	×	—	×
Natica catenoides	—	×	×	—	—
„ catena	—	×	×	—	×
„ Guillemini (?)	—	×	×	—	×
„ proxima	—	×	—	—	×
200. „ varians	—	×	×	—	—
„ hemiclausa	—	×	×	—	—
„ cirriformis	—	×	—	—	—
„ helicoides	—	×	×	—	×
„ occlusa	—	—	×	×	—
205. „ Groenlandica	—	—	—	×	×
„ clausa	—	—	×	×	×
„ multipunctata	—	×	×	—	—
Sigaretus excavatus	—	×	—	—	—
Marsenia tentaculata	—	×	—	—	×
210. Velutina lævigata	—	—	—	×	×
„ undata	—	—	—	×	×
„ virgata	—	×	—	—	—
Capulus ungaricus	—	×	×	—	×
„ obliquus	—	—	×	—	—
215. „ militaris	—	×	×	—	×
„ fallax	—	×	—	—	×
Calyptræa Chinensis	—	×	×	—	×
Tectura virginea	—	—	×	—	×
„ fulva	—	×	—	—	×
220. „ ? parvula	—	—	—	×	—
Scissurella crispata	—	×	—	—	×
Emarginula fissura	—	×	×	—	×
„ crassa	—	×	×	—	×
Cemoria Noachina	—	—	—	×	×
225. Fissurella Græca	—	×	×	—	×

ORD. *TECTIBRANCHIATA.*

	Eocene.	Miocene.	Pliocene.	Pleistocene.	Recent.
Actæon Noæ	—	×	×	—	—
„ tornatilis	—	×	×	—	×
„ subulatus	—	×	×	—	—
„ levidensis	—	×	—	—	—
230. Bulla lignaria	—	×	×	—	×
„ conulus	×	×	—	—	×
„ acuminata	×	×	—	—	×
„ cylindracea	—	×	×	—	×
„ concinna	—	×	—	—	×
235. „ truncata	—	×	—	—	×
„ Regulbiensis	—	—	—	×	×
„ Lajonkaireana	—	×	—	—	×
„ nana	—	×	—	—	—
Bullæa quadrata	—	×	—	—	×
240. „ sculpta	—	×	—	—	—
„ scabra	—	×	—	—	×
„ ventrosa	—	×	—	—	—

ORD. *CYCLOBRANCHIATA.*

	Eocene.	Miocene.	Pliocene.	Pleistocene.	Recent.
Patella vulgata	—	—	×	—	×
Chiton fascicularis	—	×	—	—	×
245. „ strigillatus	—	×	—	—	×
„ Rissoi (?)	—	×	—	—	×

ORD. *CIRRIBRANCHIATA.*

	Eocene.	Miocene.	Pliocene.	Pleistocene.	Recent.
Dentalium costatum	—	×	×	—	×?
„ entale	—	—	—	×	×?
„ bifissum	—	×	—	—	—

CLASS—PTEROPODA.

ORD. *THECOSOMATA.*

	Eocene.	Miocene.	Pliocene.	Pleistocene.	Recent.
250. Cleodora infundibulum	—	×	—	—	—

INDEX.

The Names employed are those in Roman characters, the Synonyma are in Italics.

PRINTED BY C. AND J. ADLARD,
BARTHOLOMEW CLOSE.

TAB. I.

Fig.

1. Helix rysa, *page* 4.
   - *a.* upper surface.
   - *b.* under surface.
   - *c.* front view.

2. Helix arbustorum (?), *p.* 3.
   - *a.* upper surface.
   - *b.* under surface.
   - *c.* front view.

3. Helix hispida, *p.* 2.
   - *a.* under surface.
   - *b.* upper surface.
   - *c.* front view.

4. Helix pulchella, *p.* 3.
   - *a.* upper surface.
   - *b.* under surface.
   - *c.* front view.

5. Succinea putris (?), *p.* 5.

6. Succinea oblonga, *p.* 6.

7. Limnæa peregra, *p.* 7.
   - *a.* front view.
   - *b.* back view.

8. Limnæa truncatula (?), *p.* 8.
   - *a.* var. α.
   - *b.* var. β.

Fig.

9. Limnæa palustris (?), *p.* 7.
   - *a.* front view, var. α.
   - *b.*   id.   var. β.
   - *c.* back view.
   - *d.* front view of another specimen.

10. Planorbis complanatus, *p.* 9.
    - *a.* under surface.
    - *b.* upper surface.
    - *c.* front view.

11. Planorbis spirorbis, *p.* 9.
    - *a.* upper surface.
    - *b.* under surface.
    - *c.* front view.

12. Planorbis corneus, *p.* 10.
    - *a.* under surface.
    - *b.* upper surface.
    - *c.* front view.

13. Conovulus pyramidalis, *p.* 11.
    - *a.* front view.
    - *b.* back view.

14. Conovulus myosotis (?), *p.* 12.
    - *a.* front view.
    - *b.* back view.

15. Conovulus myosotis (?), var. β.

The lines indicate the size of the specimens.

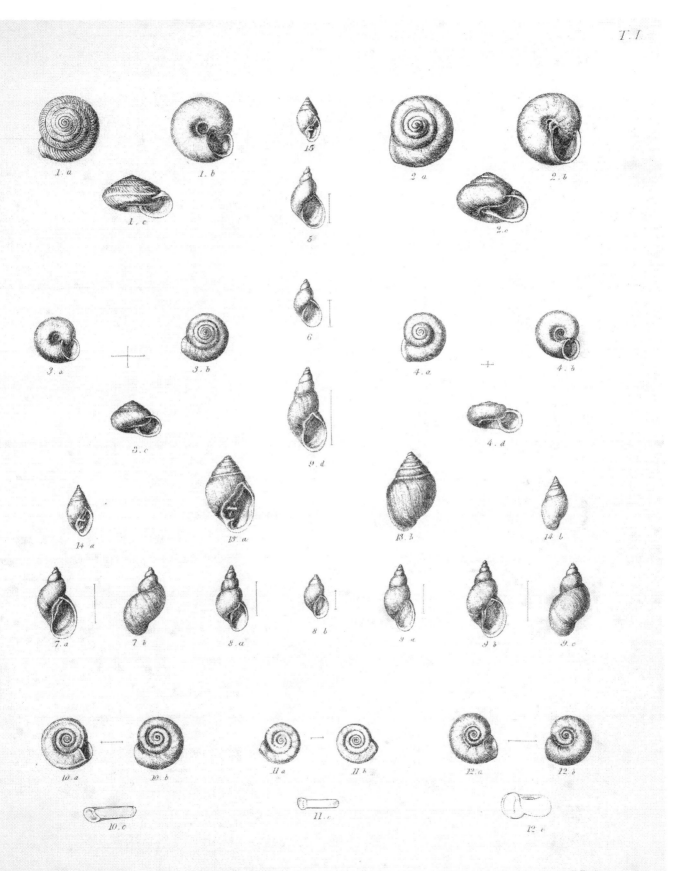

G.B.Sowerby. Jun.

TAB. II.

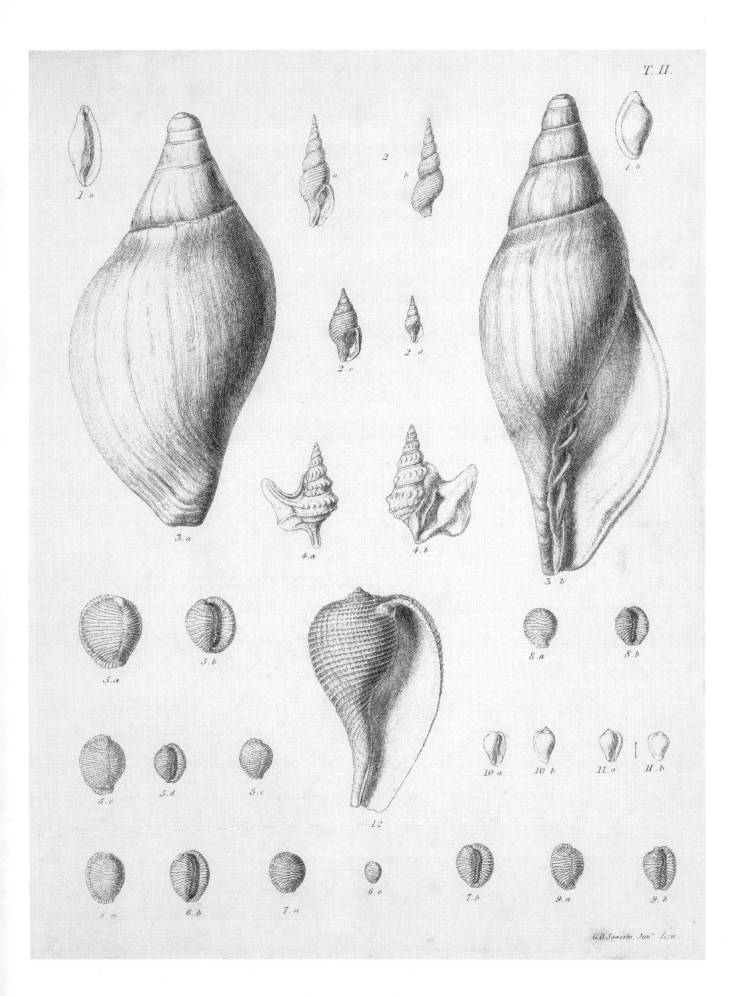

T. II.

G.B.Sowerby, Jun.r fecit.

TAB. III.

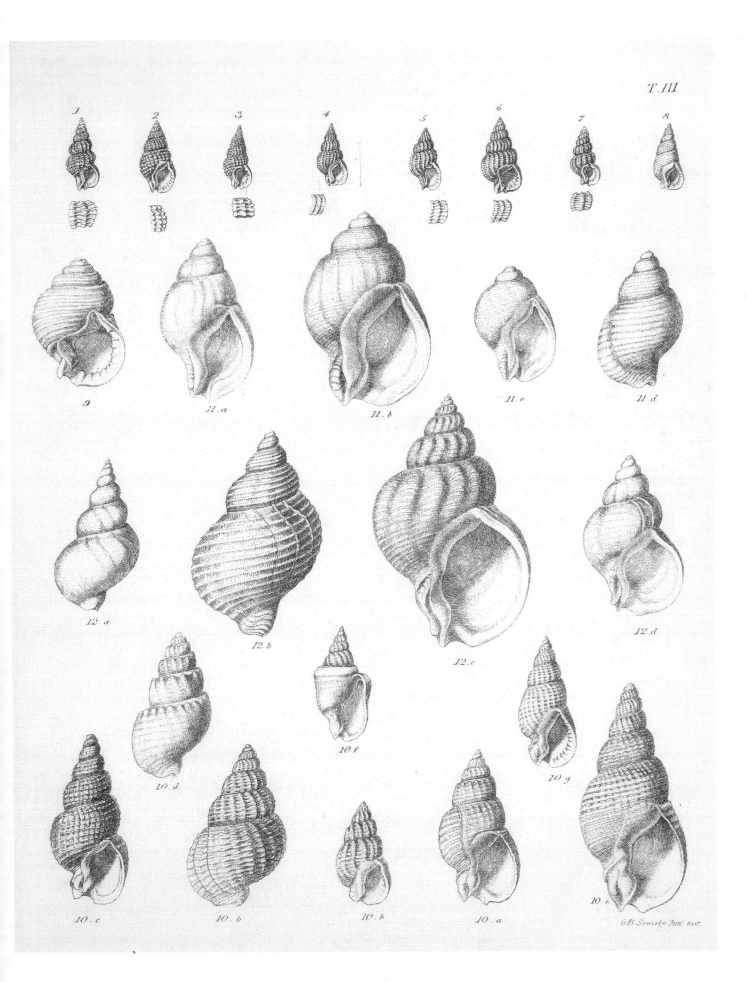

T. III

G.B. Sowerby Junr. delt.

TAB. IV.

Fig.

1. Ringicula ventricosa, *page* 22.
   - *a*. front view.
   - *b*. back view.

2. Ringicula buccinea, *p*. 22.
   - *a*. back view.
   - *b*. front view.

3. Terebra inversa, *p*. 26.
   - *a*. back view.
   - *b*. front view.

4. Terebra canalis, *p*. 26.

5. Cassidaria bicatenata, *p*. 27.
   - *a*. front view.
   - *b*. back view of another specimen.

6. Purpura lapillus, *p*. 36.
   - 6 *a*. var. crispata.
   - 6 *b*.   „   brevis.
   - 6 *c*.   „   elongata.
   - 6 *d*.   „   vulgaris.
   - 6 *e*.   „   angulata.
   - 6 *f*.   „   carinata.
   - 6 *g*.   „   imbricata.
   - 6 *h*.   „   incrassata.

7. Purpura tetragona, *p*. 38.
   - 7 *a*. var. vulgaris.
   - 7 *b*.   „   alveolata.
   - 7 *c*.   „   intermedia.
   - 7 *d*.   „   tenera.

8. Triton heptagonum, *p*. 41.

9. Murex tortuosus, *p*. 40.

1.a 1.b 2.a 2.b

4

5.a 3.b

5.a

6.a

5.b

6.e 6.b

6.d 6.c

7.a 8 7.b

9

7.c 7.d 6.f 6.g 6.h

C.B.Sowerby Jun.r del.t

TAB. V.

Fig.

1. Trophon antiquum, *page* 44.

    1 *a.* var. jugosum.

    1 *b.* „ carinatum.

    1 *c.* „ striatum.

    1 *d.* „ contrarium porrectum.

    1 *e.* „   „   rugosum.

    1 *f.* „   „   ambulacrum.

    1 *g.* „   „   striatum.

    1 *h.* „   „   angulatum.

    1 *i.* „   „   sinistrorsum.

    1 *j.* „   „   elongatum.

    1 *k.* „   „   carinatum.

2. Trophon elegans, *p.* 46.

1.d

1.a

1.k

1.e

1.f

1.b

1.j

1.g

1.h

1.e

2

1.i

G.B.Sowerby, Jun.ᵗ fecit.

TAB. VI.

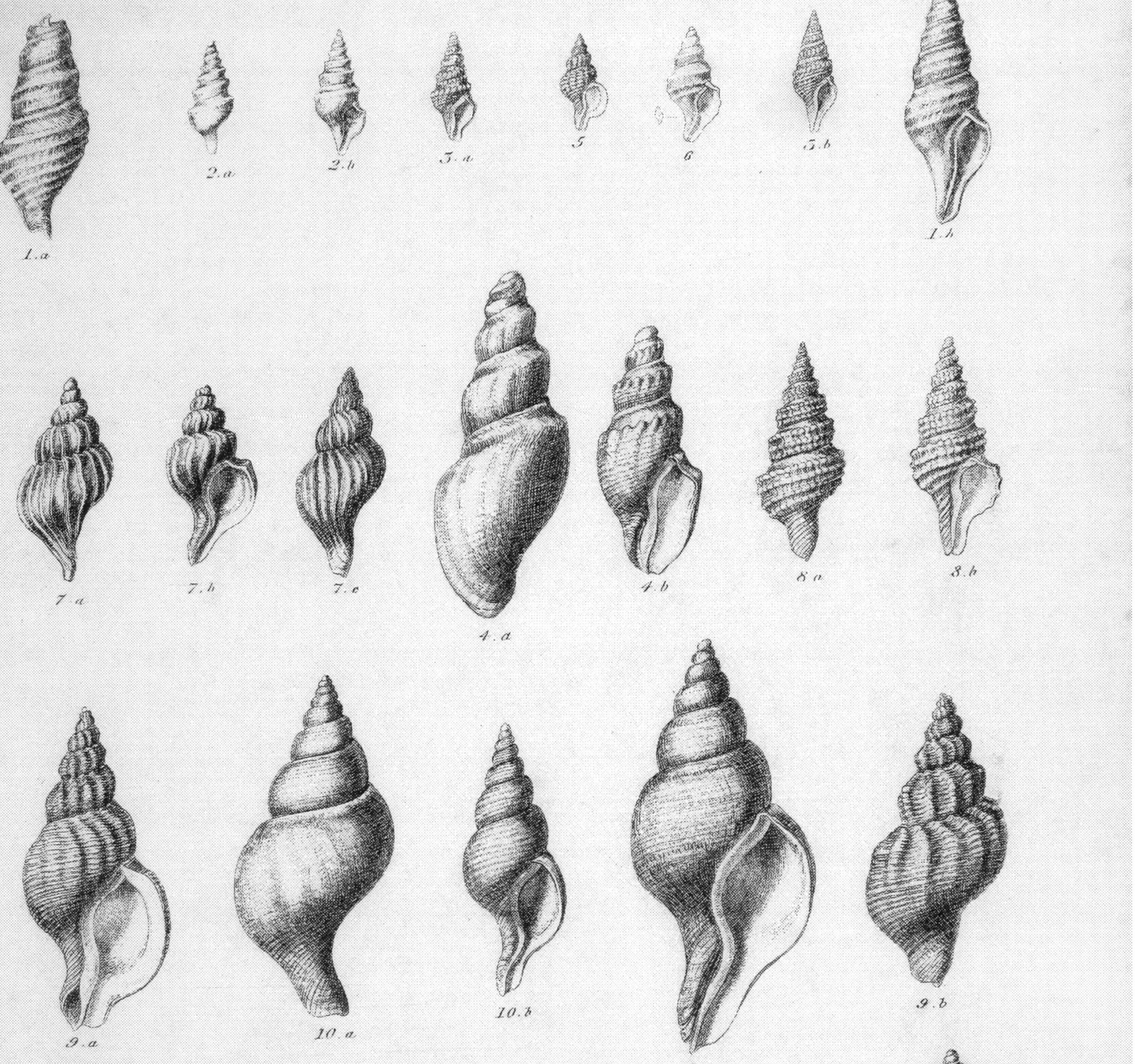

G.B. Sowerby Jun. fecit

TAB. VII.

The lines indicate the size of the specimens.

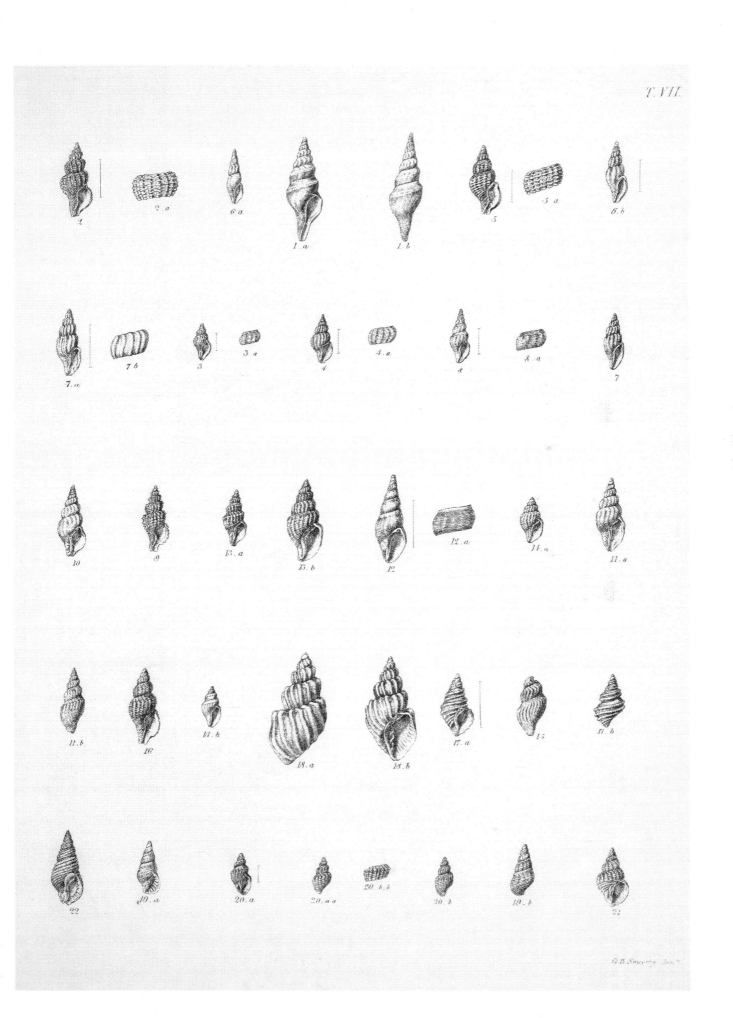

TAB. VIII.

Fig.
1. Cerithium tricinctum (?), *page* 69.
    1 *a*. front view of specimem from Mammaliferous Crag.
    1 *b*. front view of specimen from Red Crag.
2. Cerithium tricinctum (?), back view of eroded specimen from Red Crag.
3. Cerithium variculosum, *p*. 69.
    3 *a*. magnified portion.
4. Cerithium trilineatum, *p*. 70.
    4 *a*. apex magnified.
5. Cerithium tuberculare, *p*. 70.
    5 *a*. magnified portion.
    5 *b*. var. subulatum.
    5 *c*. ,, nanum.
6. Cerithium metaxa (?), *p*. 71.
    6 *a*. magnified portion.
7. Cerithium cribrarium, *p*. 71.
    7 *a*. back view.
    7 *b*. front view, another specimen.
    7 *a a*. magnified portion.
8. Cerithium adversum, *p*. 72.
    8 *a*. magnified portion.
9. Cerithium granosum, *p*. 73.
10. Cerithium perpulchrum, *p*. 72.
    10 *a*. magnified portion.

Fig.
11. Scalaria Grœnlandica, *p*. 90.
    11 *a*. back view, var. α.
    11 *b*. front view, var. β.
12. Scalaria fimbriosa, *p*. 91.
13. Scalaria hamulifera, *p*. 91.
14. Scalaria varicosa, *p*. 90.
15. Scalaria frondosa, *p*. 92.
16. Scalaria frondicula, *p*. 92.
17. Scalaria foliacea, *p*. 93.
18. Scalaria subulata, *p*. 93.
19. Scalaria clathratula, *p*. 94.
    19 *a*. front view.
    19 *b*. back view.
    a a. magnified portion.
20. Scalaria Trevellyana, *p*. 94.
21. Scalaria obtusicostata, *p*. 95.
22. Scalaria cancellata, *p*. 95.
    22 *a*. magnified portion.
23. Fossarus sulcatus, *p*. 120.
    23 *a*. front view, var. α.
    23 *b*. back view.
    23 *a a*. magnified portion.
    23 *c*. back view, var. lineolatus.
    23 *d*. front view.

The lines indicate the size of the specimens.

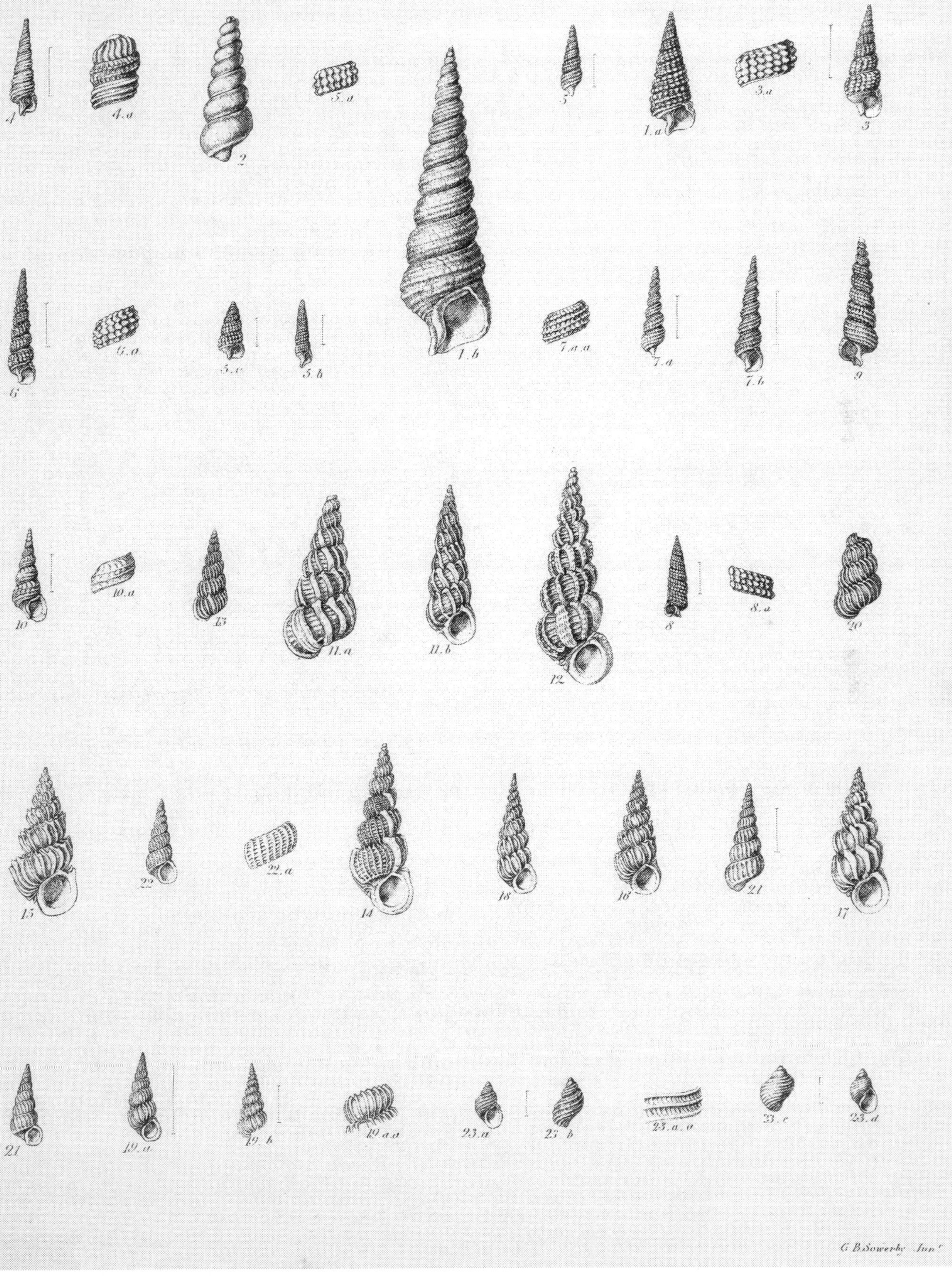

G. B. Sowerby Jun[r]

TAB. IX.

The lines indicate the size of the specimens.

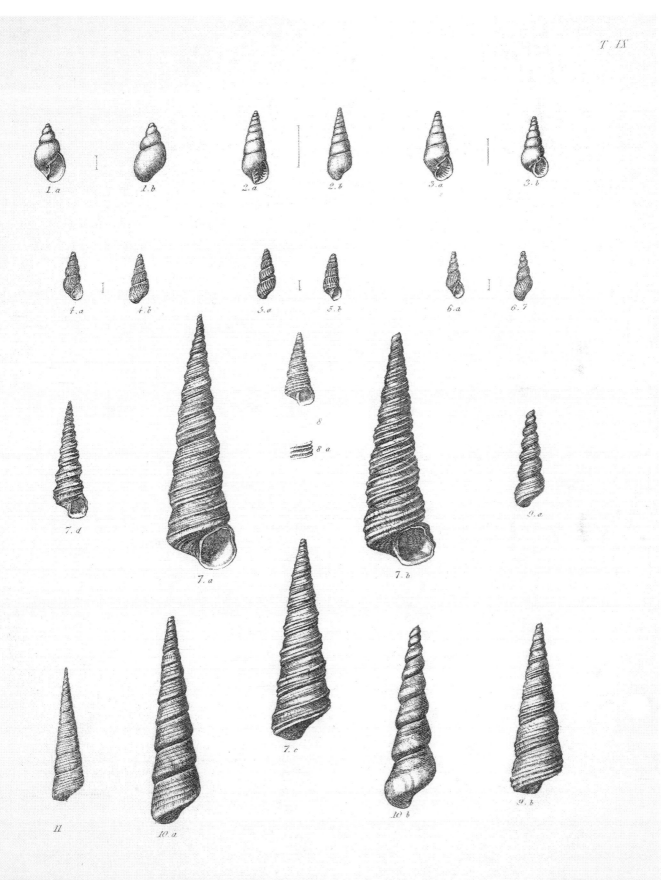

TAB. X.

The lines indicate the size of the specimens.

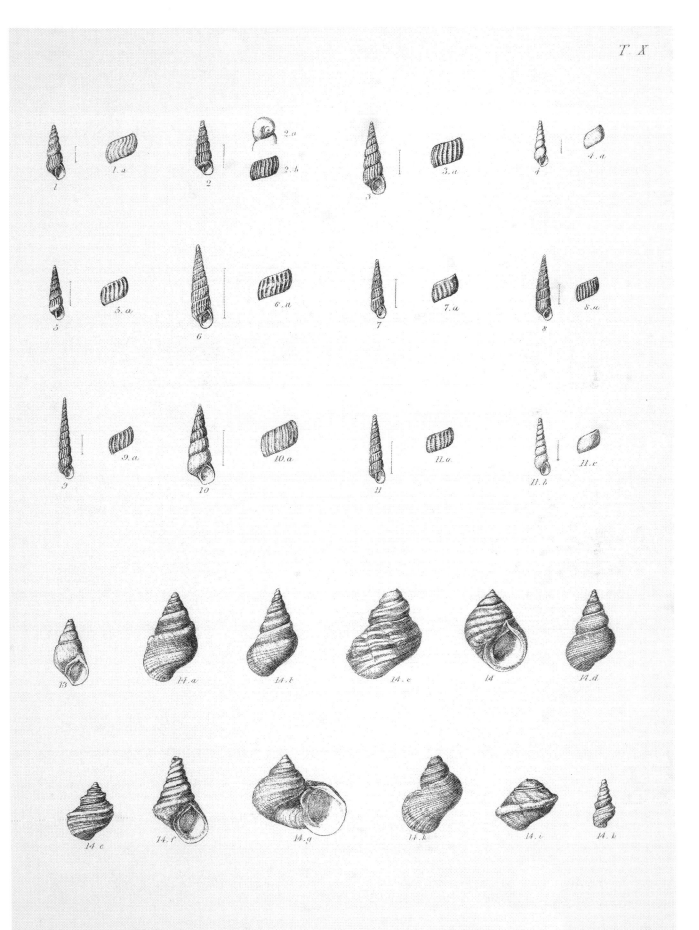

TAB. XI.

Fig.
1. Rissoa striata, *page* 100.
   a. front view.
   b. back view.

2. Paludestrina subumbilicata, *p.* 108.
   a. front view.
   b. back view.

3. Rissoa vitrea, *p.* 102.
   a. front view.
   b. back view.

4. Rissoa punctura (?), *p.* 103.
   a. front view.
   b. back view.

5. Rissoa reticulata (?), *p.* 103.
   a. front view.
   b. back view.

6. Rissoa confinis, *p.* 104.
   a. front view.
   b. back view.

Fig.
7. Rissoa Zetlandica, *p.* 101.
   a. front view.
   b. back view.

8. Rissoa supracostata, *p.* 107.
   a. front view.

9. Rissoa pulchella (?), *p.* 104.
   a. front view.
   b. back view.

10. Rissoa semicostata, *p.* 102.

11. Rissoa obsoleta, *p.* 105.
   a. front view.
   b. back view.

12. Rissoa costulata, *p.* 106.
   a. front view, var. *a.*
   b. front view, var. *β.*

13. Rissoa crassi-striata, *p.* 106.
   b. front view.
   a. back view.

The lines indicate the size of the specimens.

1.a 1.b 2.a 2.b 3.a 3.b

4.a 4.b 5.a 5.b 6.a 6.b

7.a 7.b 8 9.a 9.b

10 11.a 11.b 12.a 13.a 13.b 12.b

G.B.Sowerby jun.

TAB. XII.

The lines indicate the size of the specimens.

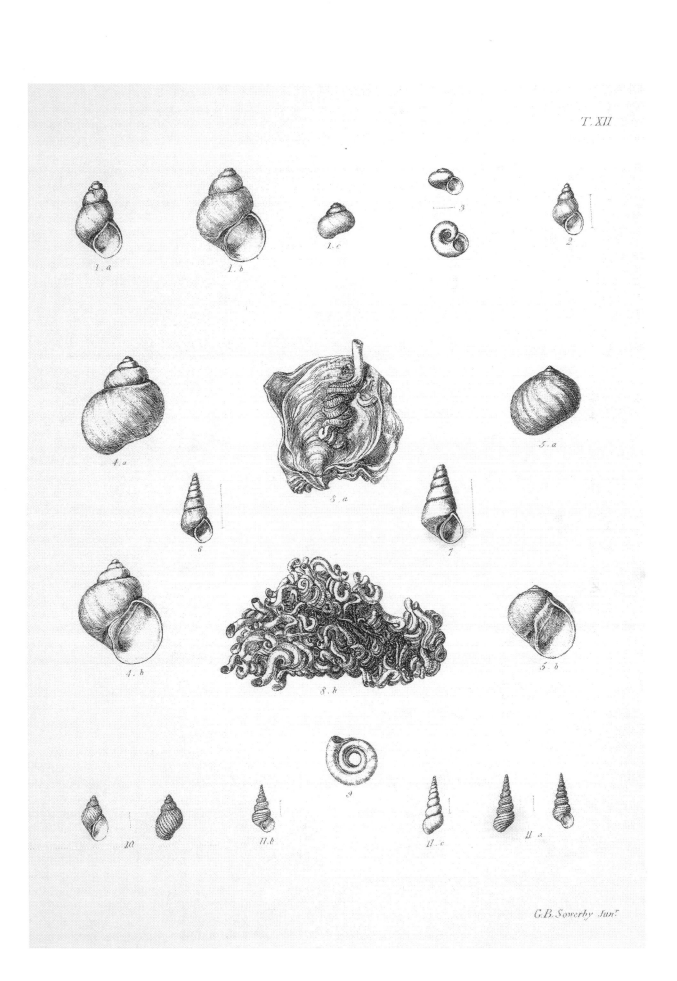

TAB. XIII.

G.B.Sowerby Jun.

TAB. XIV.

Fig.

1. Trochus Montacuti, *page* 129.
    a. front view of eroded specimen.
    b. back view of specimen with coloured stripes.

2. Trochus tumidus, *p.* 130.
    a. back view, var. conicus.
    b. front view, var. depressus.

3. Trochus Adansoni, *p.* 129.
    a. back view, var. filosa.
    b. back view, var. crenulata.
    c. front view of specimen with coloured stripes.

4. Trochus villicus, *p.* 128.
    a. back view.
    b. front view.

5. Trochus Kicksii, *p.* 130.
    a. back view.
    b. front view of elevated var.

6. Trochus tricariniferus, *p.* 132.
    a. back view, young specimen.
    b. front view, adult specimen.

7. Trochus cinerarius, *p.* 131.

8. Trochus cineroides, *p.* 131.
    a. front view, var. conicus.
    b. front view, var. umbilicatus.

9. Trochus ditropis, *p.* 133.
    a. back view.
    b. front view.

10. Trochus obconicus, *p.* 133.
    a. back view, var. β.
    b. front view, ditto.
    c. back view, var. a.
    d. front view, ditto.

The lines indicate the size of the specimens.

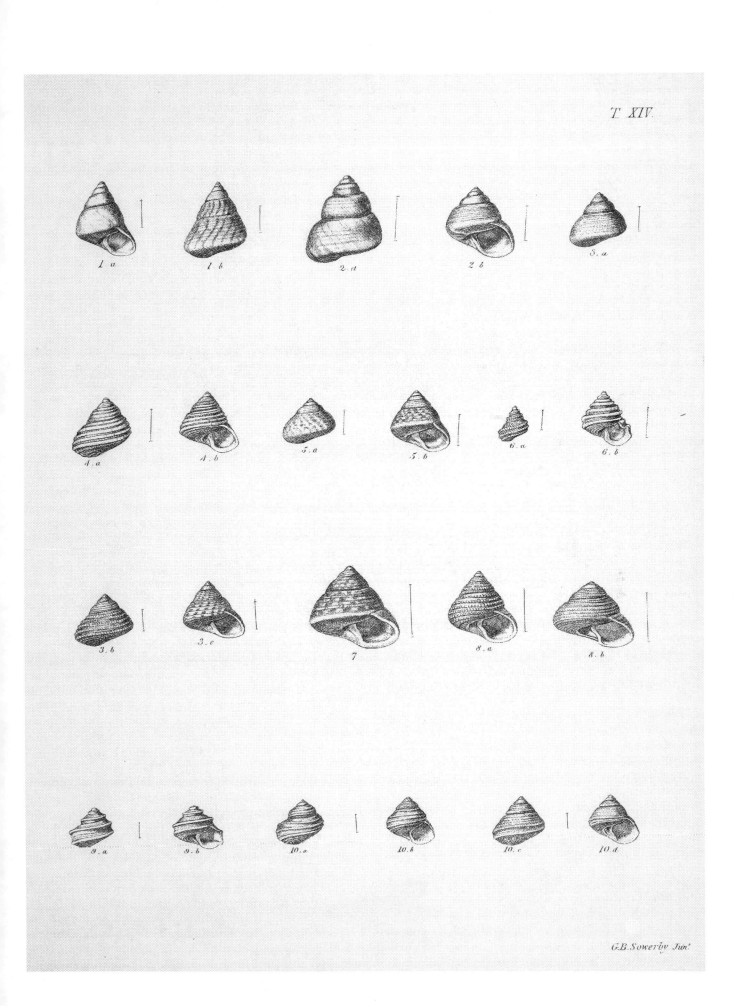

T. XIV.

G.B. Sowerby Jun.

TAB. XV.

The lines indicate the size of the specimens.

T. XV

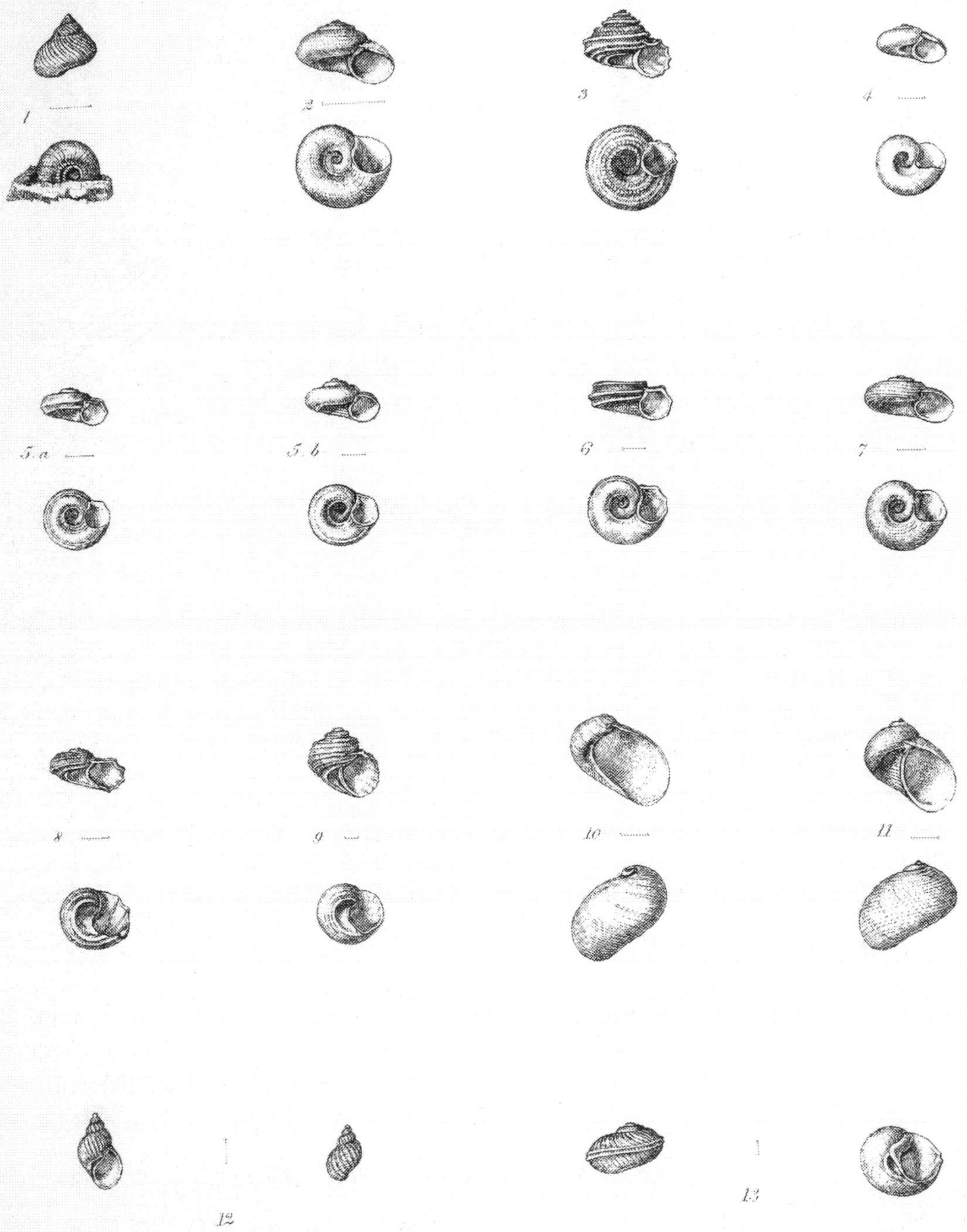

G. B. Sowerby Jun.r

TAB. XVI.

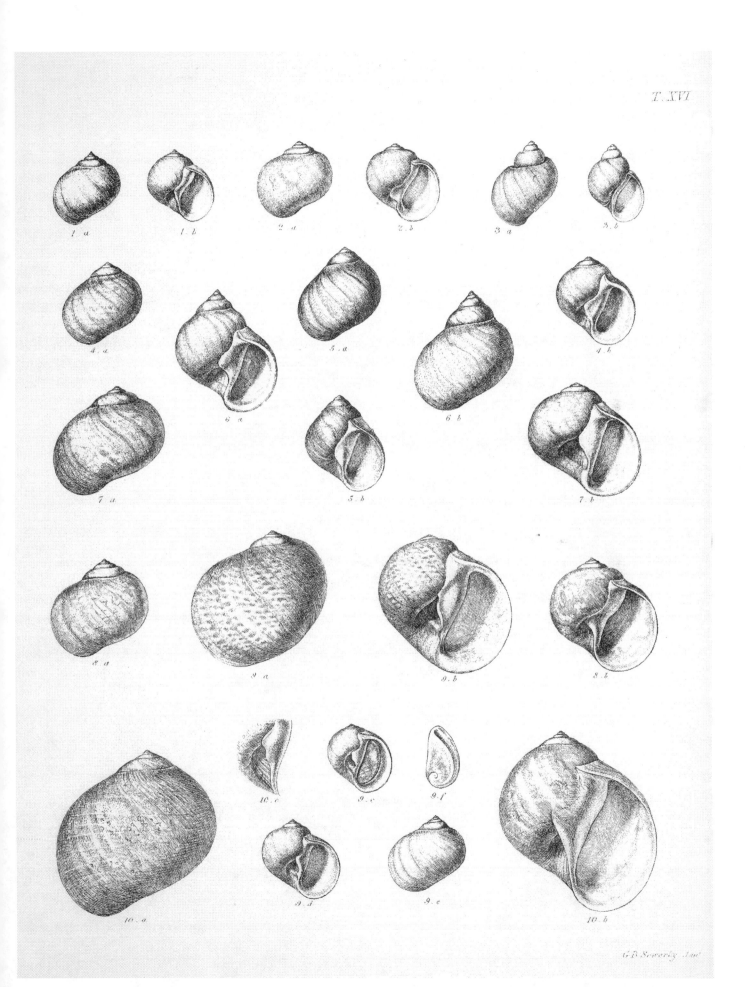

TAB. XVII.

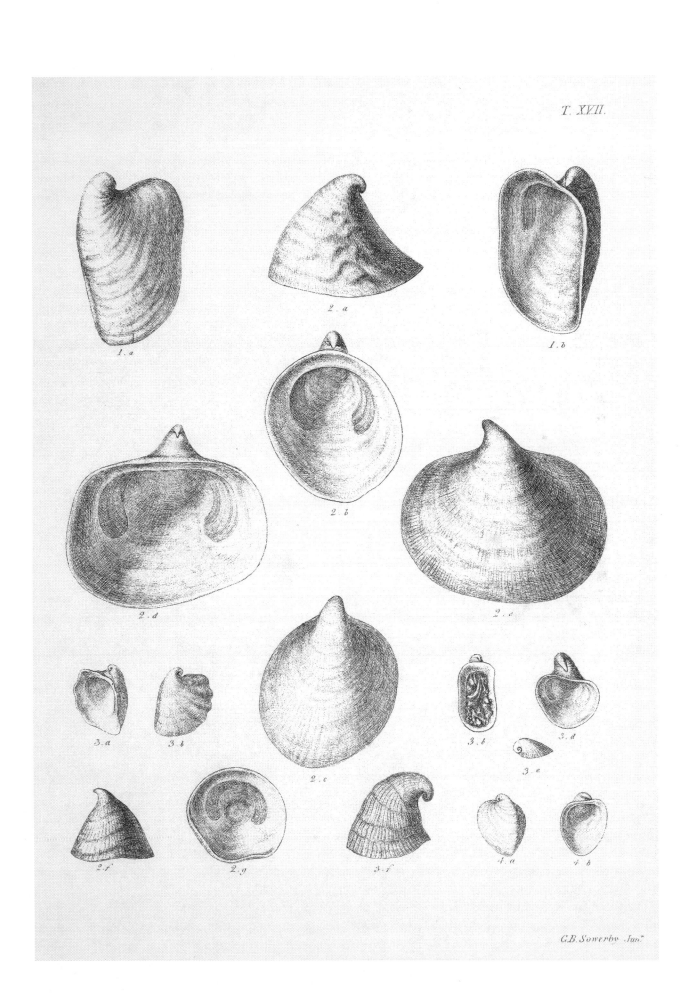

1.a

2.a

1.b

2.b

2.d

2.e

3.a   3.b

2.c

3.b   3.d

3.e

2.f   2.g   3.f   4.a   4.b

G.B.Sowerby Jun.

TAB. XVIII.

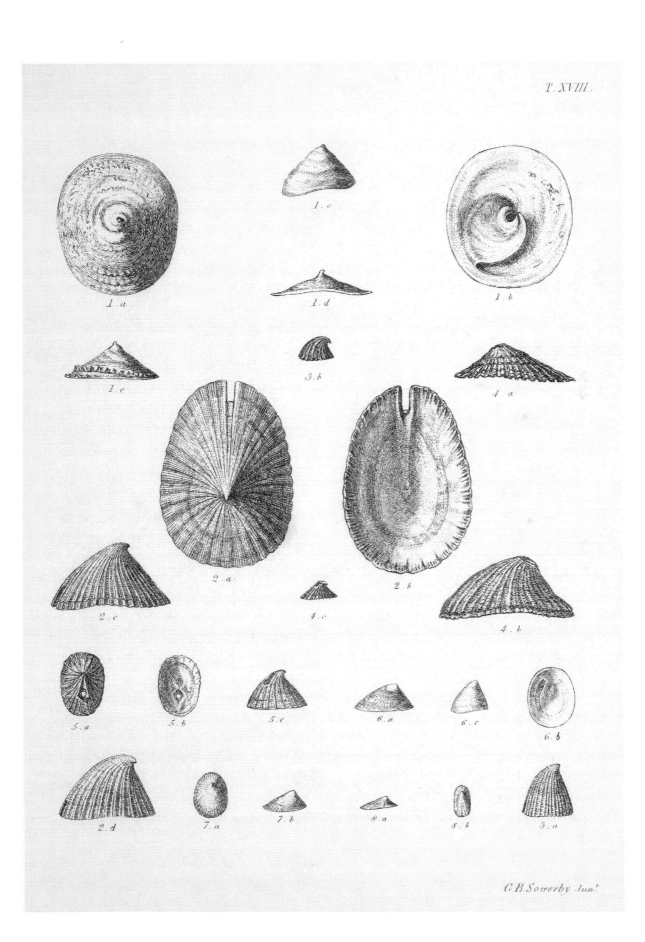

1.c

1.a

1.d

1.b

1.e

3.b

4.a

2.a

2.b

4.c

2.c

4.b

5.a

5.b

5.c

6.a

6.c

6.b

2.d

7.a

7.b

8.a

8.b

3.a

G.B.Sowerby Jun.

TAB. XIX.

The lines indicate the size of the specimens.

T. XIX

G.B.Sowerby Jun.

TAB. XX.

The lines indicate the size of the specimens.

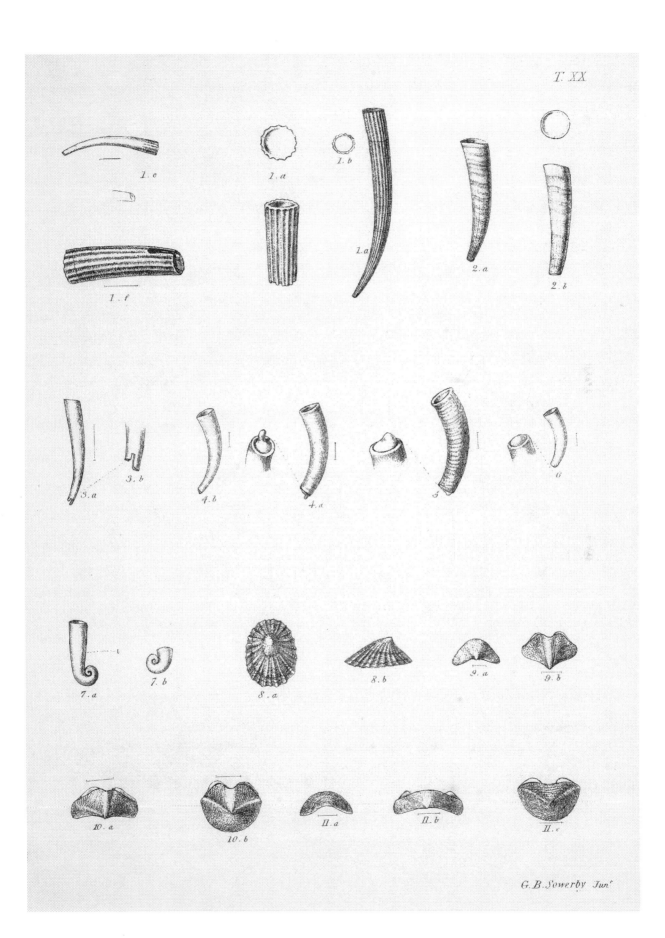

G.B.Sowerby Jun.

TAB. XXI.

The lines indicate the size of the specimens.

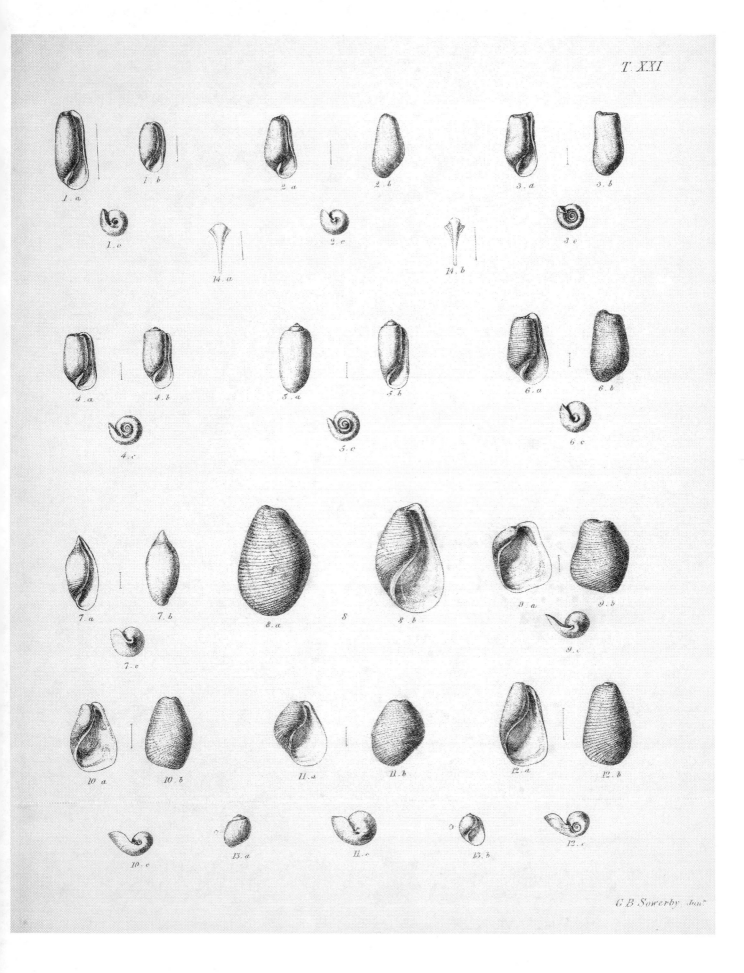

G.B. Sowerby Jun.

Printed in the United States
By Bookmasters